Freshwater Ecosystem
Ideas and Actions

Freshwater Ecosystem Ideas and Actions

Rakesh Kumar

Freshwater Ecosystem Ideas and Actions

ISBN 978-93-5111-497-0

Published in 2015 in India by

RANDOM PUBLICATIONS

4376-A/4B, Gali Murari Lal, Ansari Road
New Delhi-110 002
Phone : +9111-43580356, 011-23289044, 011-43142548
e-mail: sales@randompublications.com,
info@randompublications.com, randomexports@gmail.com

Reprinted 2019

Type Setting by : Friends Media, Delhi-110089
Printed at : Replika Press Pvt. Ltd.

Preface

Freshwater ecosystems are aquatic systems which contain drinkable water or water of almost no salt content. Freshwater resources include lakes and ponds, rivers and streams, reservoirs, wetlands, and groundwater. The world derives many benefits from these freshwater resources. They provide the majority of our drinking water resources, water resources for agriculture, industry, sanitation, as well as food including fish and shellfish. They also provide recreational opportunities and a means of transportation. In addition, freshwater ecosystems are home to numerous organisms. It has been estimated that 40 per cent of all known fish species on Earth come from freshwater ecosystems.

Despite all of their value and importance, many freshwater ecosystems are being severely damaged by human activities. The major threats to freshwater biodiversity include runoff from agricultural and urban areas, the invasion of exotic species, and the creation of dams and water diversion. Overexploitation and pollution also threaten groundwater supplies. These kinds of threats and others have already significantly impacted the biodiversity within these ecosystems.

This book examines the impacts of environmental change on freshwater ecosystems. It considers the interactions between climate change and other drivers of change including hydromorphological modification, nutrient loading, acid deposition and contamination by toxic substances. It evaluates the processes in relation to extreme events, seasonal changes in ecosystems, trends over decadal-scale time periods, mitigation strategies and ecosystem recovery.

This is an ideal book for the scientific research community, but is also accessible to masters and senior undergraduate students.

Author

Contents

1

Nature of Freshwater Ecosystems

Freshwater is naturally occurring water on the Earth's surface in ice sheets, ice caps, glaciers, bogs, ponds, lakes, rivers and streams, and underground as groundwater in aquifers and underground streams. Freshwater is generally characterized by having low concentrations of dissolved salts and other total dissolved solids. The term specifically excludes seawater and brackish water although it does include mineral rich waters such as chalybeate springs. The term "sweet water" has been used to describe freshwater in contrast to salt water

Scientifically, freshwater habitats are divided into lentic systems, which are the stillwaters including ponds, lakes, swamps and mires; lotic systems, which are running water; and groundwater which flows in rocks and aquifers. There is, in addition, a zone which bridges between groundwater and lotic systems, which is the hyporheic zone, which underlies many larger rivers and can contain substantially more water than is seen in the open channel. It may also be in direct contact with the underlying underground water.

Source of Freshwater

The source of almost all freshwater is precipitation from the atmosphere, in the form of mist, rain and snow. Freshwater falling as mist, rain or snow contains materials dissolved from the atmosphere and material from the sea and land over which the rain bearing clouds have traveled. In industrialized areas rain is typically acidic because of dissolved oxides of sulfur and nitrogen formed from burning of fossil fuels in cars, factories, trains and aircraft and from the atmospheric emissions of industry. In some cases this acid rain results in pollution of lakes and rivers.

In coastal areas freshwater may contain significant concentrations of salts derived from the sea if windy conditions have lifted drops of seawater into the rain-bearing clouds. This can give rise to elevated concentrations of sodium, chloride, magnesium and sulfate as well as many other compounds in smaller concentrations.

In desert areas, or areas with impoverished or dusty soils, rain-bearing winds can pick up sand and dust and this can be deposited elsewhere in precipitation and causing the freshwater flow to be measurably contaminated both by insoluble solids but also by the soluble components of those soils. Significant quantities of iron may be transported in this way including the well-documented transfer of iron-rich rainfall falling in Brazil derived from sand-storms in the Sahara in north Africa.

Distribution

Water is a critical issue for the survival of all living organisms. Some can use salt water but many organisms including the great majority of higher plants and most mammals must have access to freshwater to live. Some terrestrial mammals, especially desert rodents appear to survive without drinking but they do generate water through the metabolism of cereal seeds and they also have mechanisms to conserve water to the maximum degree.

Out of all the water on Earth, salt water in oceans, seas and saline groundwater make up about 97% of it. Only 2.5-2.75% is freshwater, including 1.75-2% frozen in glaciers, ice and snow, 0.7-0.8% as fresh groundwater and soil moisture, and less than 0.01% of it as surface water in lakes, swamps and rivers.

Freshwater lakes contain about 87% of this fresh surface water, including 29% in the African Great Lakes, 23% in Lake Baikal in Russia, 21% in the North American Great Lakes, and 14% in other lakes. Swamps have most of the balance with only a small amount in rivers, most notably the Amazon River. The atmosphere contains 0.04% water. In areas with no freshwater on the ground surface, freshwater derived from precipitation may, because of its lower density, overlie saline ground water in lenses or layers. Most of the world's freshwater is frozen in ice sheets. Many areas suffer from lack of distribution of freshwater, such as deserts.

Freshwater as a Resource

An important concern for hydrological ecosystems is securing minimum streamflow, especially preserving and restoring instream water allocations. Freshwater is an important natural resource necessary for the survival of all ecosystems. The use of water by humans for activities such as irrigation and industrial applications can have adverse impacts on down-stream ecosystems. Chemical contamination of freshwater can also seriously damage eco-systems.

Pollution from human activity, including oil spills, also presents a problem for freshwater resources. The largest petroleum spill that has ever occurred in freshwater was caused by a Royal Dutch Shell tank ship in Magdalena, Argentina, on January 15, 1999, polluting the environment, drinkable water, plants and animals.

Fresh and unpolluted water accounts for 0.003% of total water available globally.

Agriculture

Changing landscape for the use of agriculture has a great effect on the flow of freshwater. Changes in landscape by the removal of trees and soils changes the flow of freshwater in the local environment and also affects the cycle of freshwater. As a result more freshwater is stored in the soil which benefits agriculture. However, since agriculture is the human activity that consumes the most freshwater, this can put a severe strain on local freshwater resources resulting in the destruction of local ecosystems. In Australia, over-abstraction of freshwater for intensive irrigation activities has caused 33% of the land area to be at risk of salination. With regards to agriculture, the World Bank targets food production and water management as an increasingly global issue that will foster debate.

Limited Resource

Freshwater is a renewable and changeable, but limited natural resource. Freshwater can only be renewed through the process of the water cycle, where water from seas, lakes, rivers, and dams evaporates, forms clouds, and returns to water sources as precipitation. However, if more freshwater is consumed through human activities than is restored by nature, the result is that the quantity of freshwater available in lakes, rivers, dams and underground waters is reduced which can cause serious damage to the surrounding environment.

Freshwater Withdrawal

Freshwater withdrawal is the quantity of water removed from available sources for use in any purpose. Water drawn off is not necessarily entirely consumed and some portion may be returned for further use downstream.

Causes of Limited Freshwater

There are many causes of the apparent decrease[citation needed] in our freshwater supply. Principal amongst these is the increase in population through increasing life expectancy, the increase in per capita water use and the desire of many people to live in warm climates that have naturally low levels of freshwater resources.Climate change is also likely to change the availability and distribution of freshwater across the planet:

> "If global warming continues to melt glaciers in the polar regions, as expected, the supply of freshwater may actually decrease. First, freshwater from the melting glaciers will mingle with salt water in the oceans and become too salty to drink. Second, the increased ocean volume will cause sea levels to rise, contaminating freshwater sources along coastal regions with seawater".

The World Bank adds that the response by freshwater ecosystems to a changing climate can be described in terms of three interrelated components: water quality, water quantity or volume, and water timing. A change in one often leads to shifts in the others as well. Water pollution and subsequent eutrophication also reduces the availability of freshwater. Also, there is an uneven distribution of freshwater. While some countries have an abundant supply of freshwater, others do not have as much. For example, Canada has 20% of the world's freshwater supply, while India has only 10% of the world's freshwater supply; what more India's population is more than 30 times larger than that of Canada. A reason for the uneven distribution of freshwater supply may be the differences in climate. For example, in some countries in Africa, the frequent lack of rain has led to insufficient water supply for irrigation.

Choices in the Use of Freshwater

With one in eight people in the world not having access to safe water it is important to use this resource in a prudent manner. Making the best use of water on a local basis probably provides the best solution. Local communities need to plan their use of freshwater and should be made aware of how certain crops and animals use water.

Freshwater Ecosystems

Freshwater ecosystems are a subset of Earth's aquatic ecosystems. They include lakes and ponds, rivers, streams and springs, and wetlands. They can be contrasted with marine ecosystems, which have a larger salt content. Freshwater habitats can be classified by different factors, including temperature, light penetration, and vegetation.

Freshwater ecosystems can be divided into lentic ecosystems (still water) and lotic ecosystems (flowing water).

Limnology (and its branch freshwater biology) is a study about freshwater ecosystems. It is a part of hydrobiology.

Original efforts to understand and monitor freshwater ecosystems were spurred on by threats to human health (ex. Cholera outbreaks due to sewage contamination). Early monitoring focussed on chemical indicators, then bacteria, and finally algae, fungi and protozoa. A new type of monitoring involves differing groups of organisms (macroinvertebrates, macrophytes and fish) and the stream conditions associated with them.

Current biomonitering techniques focus mainly on community structure or biochemical oxygen demand. Responses are measured by behavioural changes, altered rates of growth, reproduction or mortality. Macroinvertebrates are most often used in these models because of well known taxonomy, ease of collection, sensitivity to a range of stressors, and their overall value to the ecosystem. Most of these measurements are difficult to extrapolate on a large scale however.

The use of reference sites is common when assessing what a healthy freshwater ecosystem should "look like". Reference sites are easier to reconstruct in standing water than moving water. Preserved indicators such as diatom valves, macrophyte pollen, insect chitin and fish scales can be used to establish a reference ecosystem representative of a time before large scale human disturbance.

Common chemical stresses on freshwater ecosystem health include acidification, eutrophication and copper and pesticide contamination.

CHARACTERISTICS OF AQUATIC ECOSYSTEMS

The types of organisms in an aquatic ecosystem are mainly determined by the water's *salinity*—the amount of dissolved salts the water contains. As a result, aquatic ecosystems are divided into freshwater ecosystems and marine ecosystems.

Freshwater ecosystems include the sluggish waters of lakes and ponds and the moving waters of rivers and streams. They also include areas where land, known as a *wetland,* is periodically underwater. Marine ecosystems include the diverse coastal areas of marshes, swamps, and coral reefs as well as the deep, vast oceans.

Factors such as temperature, sunlight, oxygen, and nutrients determine which organisms live in which areas of the water. For instance, sunlight reaches only a certain distance below the surface of the water, so most photosynthetic organisms live on or near the surface.

Aquatic ecosystems contain several types of organisms that are grouped by their location and by their adaptations. Three groups of aquatic organisms include plankton, nekton, and benthos. *Plankton* are the organisms that float near the surface of the water. Two types of plankton are microscopic plants called *phy-toplankton,* and microscopic animals called *zooplankton.* Phyto-plankton produce most of the food for an aquatic ecosystem. *Nekton* are free-swimming organisms, such as fish, turtles, and whales. *Benthos* are bottom-dwelling organisms, such as mussels, worms, and barnacles. Many benthic organisms live attached to hard surfaces. Decomposers, organisms that break down dead organisms, are also a type of aquatic organism.

LAKES AND PONDS

Lakes, ponds, wetlands, rivers, and streams make up the various types of freshwater ecosystems. Lakes, ponds, and wetlands can form naturally where groundwater reaches the Earth's surface. Beavers can also create ponds by damming up streams. Humans intentionally create artificial lakes by damming flowing rivers and streams to use them for power, irrigation, water storage, and recreation.

Life in a Lake

Lakes and ponds can be structured into horizontal and vertical zones. In the nutrient-rich *littoral zone* near the shore, aquatic life is diverse and abundant.

Plants, such as cattails and reeds, are rooted in the mud underwater, and their upper leaves and stems emerge above the water. Plants that have floating leaves, such as pond lilies, are rooted here also. Farther out from the shore, in the open water, plants, algae, and some bacteria capture solar energy to make their own food during *photosynthesis.* The types of organisms present in a pond or lake ecosystem depend on the amount of sunlight available.

Some bodies of freshwater have areas so deep that there is too little light for photosynthesis. Bacteria live in the deep areas of the freshwater to decompose dead plants and animals that drift down from the land and water above. Fish adapted to cooler, darker water also live there. Eventually, dead and decaying organisms reach the *benthic zone,* the bottom of a pond or lake, which is inhabited by decomposers, insect larvae, and clams.

Animals that live in lakes and ponds have adaptations that help them obtain what they need to survive. Water beetles use the hairs under their bodies to trap surface air so that they can breathe during their dives for food. Whiskers help catfish sense food as they swim over dark lake bottoms. In regions where lakes partially freeze in winter, amphibians burrow into the littoral mud to avoid freezing temperatures.

How Nutrients Affect Lakes

Eutrophication is an increase in the amount of nutrients in an aquatic ecosystem. A lake that has a large amount of plant growth due to nutrients, is known as a *eutrophic lake.* As the amount of plants and algae grows, the number of bacteria feeding on the decaying organisms also grows. These bacteria use the oxygen dissolved in the lake's waters. Eventually, the reduced amount of oxygen kills oxygen-loving organisms. Lakes naturally become eutrophic over a long period of time. However, eutrophication can be accelerated by runoff. Runoff is precipitation, such as rain, that can carry sewage, fertilizers, or animal wastes from land into bodies of water.

Freshwater Wetlands

Freshwater wetlands are areas of land that are covered with freshwater for at least part of the year. The two main types of freshwater wetlands are marshes and swamps. *Marshes* contain non-woody plants, such as cattails, while *swamps* are dominated by woody plants, such as trees and shrubs.

Wetlands perform several important environmental functions. Wetlands act as filters or sponges because they absorb and remove pollutants from the water that flows through them. Therefore, wetlands improve the water quality of lakes, rivers, and reservoirs downstream. Wetlands also control flooding by absorbing extra water when rivers overflow, which protects farms and urban and residential areas from damage. Many of the freshwater game fish caught in the United States each year use the wetlands for feeding and spawning. In addition, these areas provide a home for native and migratory wildlife. Wetland vegetation also traps carbon that would otherwise be released as carbon dioxide, which may be linked to rising atmospheric temperatures. Some wetlands are used to produce many commercially important products, such as cranberries.

Table 1. Environmental Functions of Wetlands

- trapping and filtering sediments, nutrients, and pollutants, which keep these materials from entering lakes, reservoirs, and oceans
- reducing the likelihood of a flood, protecting agriculture, roads, buildings, and human health and safety
- buffering shorelines against erosion
- providing spawning grounds and habitat for commercially important fish and shellfish
- providing habitat for rare, threatened, and endangered plants and animals
- providing recreational areas for activities such as fishing, bird-watching, hiking, canoeing, photography, and painting

Marshes

*M*ost freshwater wetlands are located in the southeastern United States. The Florida Everglades is the largest freshwater wetland in the United States. Freshwater marshes tend to occur on low, flat lands and have little water movement. In shallow waters, plants such as reeds, rushes, and cattails root themselves in the rich bottom sediments. The leaves of these and other plants stick out above the surface of the water year-round.

The benthic zones of marshes are nutrient rich and contain plants, numerous types of decomposers, and scavengers. Waterfowl, such as grebes and ducks, have flat beaks adapted for sifting through the water for fish and insects. Water birds, such as herons, have spearlike beaks that they use to grasp small fish and to probe for frogs buried in the mud. Marshes also attract many migratory birds from temperate and tropical habitats.

There are several kinds of marshes, each of which is characterized by its salinity. Brackish marshes have slightly salty water, while salt marshes contain saltier water. In each marsh type, organisms are adapted to live within the ecosystem's range of salinity.

Swamps

Swamps occur on flat, poorly drained land, often near streams. Swamps are dominated by woody shrubs or water-loving trees, depending on the latitude and climate in which the swamps are located. Mangrove swamps occur in warm climates near the ocean, so their water is salty. Freshwater swamps are the ideal habitat for many amphibians, such as frogs and salamanders, because of the continuously moist environment. Swamps also attract birds, such as wood ducks, that nest in hollow trees near or over the water.

Human Impact on Wetlands

Wetlands were previously considered to be wastelands that provide breeding grounds for insects. Therefore, many have been drained, filled, and cleared for farms or residential and commercial development. For example, the Florida Everglades once covered 8 million acres of south Florida but it now covers less than 2 million acres. The important role of wetlands as purifiers of wastewater and in flood prevention is now recognized. Wetlands are vitally important as habitats for wildlife. Law and the federal government protect many wetlands, and most states now prohibit the destruction of certain wetlands.

Rivers

Many rivers originate from snow melt in mountains. At its headwaters, a river is usually cold and full of oxygen and runs swiftly through a shallow riverbed. As a river flows down a mountain, a river may broaden, become warmer, wider, and slower, and decrease in oxygen. A river changes with the land and the climate through which it flows. Runoff, for example, may wash nutrients and sediment from the surrounding land into a river. These materials affect the growth and health of the organisms in the river.

Life in a River

Near the churning headwaters, mosses anchor themselves to rocks by using rootlike structures called *rhizoids*. Plankton do not live in the headwaters because the current is too strong for them to float. However, trout and minnows are adapted to the cold, oxygen-rich headwaters. Trout are powerful swimmers and have streamlined bodies that present little resistance to the strong current. Farther downstream, plankton can float in the warmer, calmer waters. Other plants, such as the crowfoot, set roots down in the river's rich sediment. The leaves of some plants, such as the arrowhead, will vary in shape according to the strength of a river's current. Fish such as catfish and carp also live in the calmer waters.

River Ecosystem

The organisms in the riparian zone respond to changes in river channel location and patterns of flow. The ecosystem of rivers is generally described by the River continuum concept, which has some additions and refinements to allow for spatial (dams, waterfalls) and temporal (extensive flooding). The basic idea is that the river can be described as a system that is continuously changing along its length in the physical parameters, the availability of food particles and the composition of the ecosystem. The food (energy) that is the leftover of the upstream part is being utilized downstream.

The general pattern is that the first order streams contain particulate matter (decaying leaves from the surrounding forests), which is processed there by shredders like Plecoptera larvae. The leftovers of the shredders are utilized by collectors, such as Hydropsychidae, and further downstream algae that create the primary production become the main foodsource of the organisms. All changes are gradual and the distribution of each species can be described as a normal curve with the highest density where the conditions are optimal. In rivers succession is virtually absent and the composition of the ecosystem stays fixed in time.

Uses of Rivers

Rivers have been used as a source of water, for obtaining food, for transport, as a defensive measure, as a source of hydropower to drive machinery, for bathing, and as a means of disposing of waste.

Rivers have been used for navigation for thousands of years. The earliest evidence of navigation is found in the Indus Valley Civilization, which existed in northwestern Pakistan around 3300 BC. Riverine navigation provides a cheap means of transport, and is still used extensively on most major rivers of the world like the Amazon, the Ganges, the Nile, the Mississippi, and the Indus. Since river boats are often not regulated, they contribute a large amount to global greenhouse gas emissions, and to local cancer due to inhaling of particulates emitted by the transports.

In some heavily forested regions such as Scandinavia and Canada, lumberjacks use the river to float felled trees downstream to lumber camps for further processing, saving much effort and cost by transporting the huge heavy logs by natural means.

Rivers have been a source of food since pre-history. They can provide a rich source of fish and other edible aquatic life, and are a major source of freshwater, which can be used for drinking and irrigation. It is therefore no surprise to find most of the major cities of the world situated on the banks of rivers. Rivers help to determine the urban form of cities and neighbourhoods and their corridors often present opportunities for urban renewal through the development of foreshoreways such as riverwalks. Rivers also provide an easy means of disposing of waste-water and, in much of the less developed world, other wastes.

Fast flowing rivers and waterfalls are widely used as sources of energy, via watermills and hydroelectric plants. Evidence of watermills shows them in use for many hundreds of years such as in Orkney at Dounby Click Mill. Prior to the invention of steam power, water-mills for grinding cereals and for processing wool and other textiles were common across Europe. In the 1890s the first machines to generate power from river water were established at places such as Cragside in Northumberland and in recent decades there has been a significant increase in the development of large scale power generation from water, especially in wet mountainous regions such as Norway.

The coarse sediments, gravel, and sand, generated and moved by rivers are extensively used in construction. In parts of the world this can generate extensive new lake habitats as gravel pits re-fill with water. In other circumstances it can destabilise the river bed and the course of the river and cause severe damage to spawning fish populations which rely on stable gravel formations for egg laying.

In upland rivers, rapids with whitewater or even waterfalls occur. Rapids are often used for recreation, such as whitewater kayaking.

Rivers have been important in determining political boundaries and defending countries. For example, the Danube was a long-standing border of the Roman Empire, and today it forms most of the border between Bulgaria and Romania. The Mississippi in North America and the Rhine in Europe are major east-west boundaries in those continents. The Orange and Limpopo Rivers in southern Africa form the boundaries between provinces and countries along their routes.

Rivers in Danger

Industries use river water in manufacturing processes and as receptacles for waste. For many years, people have used rivers to dispose of their sewage and garbage. These practices have polluted rivers with toxins, which have killed river organisms and made river fish inedible. Today, runoff from the land puts pesticides and other poisons into rivers and coats riverbeds with toxic sediments. Dams also alter the ecosystems in and around a river.

References

Friberg, N, et.al. (2011). Biomonitoring of Human Impacts in Freshwater Ecosystems: The Good, the Bad and the Ugly. *Advances in Ecological Research,* 1-68.

Leathwick J.R.; Collier K.; Chadderton L. (2007) "Identifying freshwater ecosystems with nationally important natural heritage values: development of a biogeographic framework". *Science for Conservation* 274. p. 30. Department of Conservation New Zealand.

Ricciardi, A., & Rasmussen, J. B. (1999). Extinction Rates of North American Freshwater Fauna. *Conservation Biology,* 13 (5), 1220-1222.

XU, F.L., Tao, S., Dawson, R. W., Pen-gang, L., & Jun, C. (2001). Lake Ecosystem Health Assessment: Indicators and Methods. *Water Research,* 3157-3167.

2

State of Freshwater Resources

Global freshwater use is estimated to expand 10% from 2000 to 2010, down from a per decade rate of about 20% between 1960 and 2000. These rates reflect population growth, economic development, and changes in water use efficiency. Projections that this trend will continue have a high degree of certainty. Contemporary water withdrawal is approximately 3,600 cubic kilometers per year globally or 25% of the continental runoff to which the majority of the population has access during the year. If dedicated instream uses for navigation, waste processing, and habitat management are considered, humans then use and regulate over 40% of renewable accessible supplies.

Regional variations from differential development pressures and efficiency changes during 1960–2000 produced increases in water use of 15– 32% per decade. Four out of every five people live downstream of, and are served by, renewable freshwater services, representing 75% of the total supply. Because the distribution of freshwater is uneven in space and time, more than 1 billion people live under hydrologic conditions that generate no appreciable supply of renewable freshwater. An additional 4 billion (65% of world population) is served by only 50% of total annual renewable runoff that is positioned in dry to only moderately wet conditions, with concomitant pressure on that resource base. Only about 15% live with relative water abundance. Forest and mountain ecosystems serve as source areas for the largest amounts of renewable freshwater supply—57% and 28% of total runoff, respectively. These ecosystems each provide renewable water supplies to at least 4 billion people, or two thirds of the global population. Cultivated and urban ecosystems generate only 16% and 0.2%, respectively, of global runoff, but because of their close proximity to human settlements, they serve 4–5 billion people. Such proximity is also associated with nutrient and industrial water pollution.

From 5% to possibly 25% of global freshwater use exceeds long-term accessible supply. Overuse implies delivery of freshwater services through engineered water transfers or nonrenewable groundwater supplies that are currently being depleted. Much of this water is used for irrigation with irretrievable losses in water-scarce regions. All continents record overuse. In the relatively dry Middle East and North Africa, non-sustainable use is exacerbated, with current rates of freshwater use equivalent to 115% of total renewable runoff. In addition, possibly one third of all withdrawals come from nonrenewable sources, a condition driven mainly by irrigation demand. Crop production requires enormous quantities of freshwater; consequently, many countries that aim at self-sufficiency in food production have entrenched patterns of water scarcity. Alternatively, crops can be traded on global food markets, with some countries accruing substantial benefits from importing "virtual water" that would otherwise be required domestically to irrigate crops.

The water requirements of aquatic ecosystems in the context of expanding human freshwater use results in competition for the same resources. Changes in flow regime, transport of sediments and chemical pollutants, modification of habitat, and disruption of migration routes of aquatic biota are some of the key consequences of this competition. In many parts of the world, competition for freshwater has produced impacts that fully extend to the coastal zone, with effects including oxygen depletion, coastal erosion, and harmful algal blooms. Through consumptive use and interbasin transfers, several of the world's largest rivers (the Nile, the Yellow, and the Colorado in the United States) have been transformed into highly stabilized and in some cases seasonally nondischarging river channels.

The supply of freshwater continues to be reduced by severe pollution from anthropogenic sources in many parts of the world. Over the past half-century, there has been an accelerated release of artificial chemicals into the environment. Inorganic nitrogen pollution of inland waterways, for example, has increased substantially, with nitrogen loads transported by the global system of rivers rising more than twofold over the preindustrial state.

Increases of more than tenfold are recorded across many industrialized regions of the world. Many anthropogenic chemicals are long-lived and transformed into by-products whose behaviors, synergies, and impacts are for the most part unknown as yet. As a consequence of pollution, the ability of ecosystems to provide clean and reliable sources of freshwater is impaired. Severe deterioration in the quality of freshwater is magnified in cultivated and urban systems (high use, high pollution sources) and dryland systems (high demand for flow regulation, absence of dilution potential).

The demand for reliable sources of freshwater and flood control has encouraged engineering practices that have compromised the sustainability of inland water systems and their provision of freshwater services. Prolific dam-building (45,000 large dams and possibly 800,000 smaller ones) has generated both positive and negative effects. Positive effects on human well-being have included flow stabilization for irrigation, flood control, drinking water, and hydroelectricity. Negative effects have included fragmentation and destruction of habitat, loss of species, health issues associated with stagnant water, and loss of sediments and nutrients destined to support coastal ecosystems and fisheries.

Water scarcity is a globally significant and accelerating condition for 1–2 billion people worldwide, leading to problems with food production, human health, and economic development. A high degree of uncertainty surrounds these estimates, and defining water scarcity merits substantial further analysis in order to support sound water policy formulation and management. Rates of increase in a key water scarcity measure—water use relative to accessible supply—from 1960 to present averaged nearly 20% per decade globally, with values of 15% to more than 30% per decade for individual continents. Inequalities in level of economic development, education, and governance result in differences in coping capacity for water scarcity.

The annual burden of disease from inadequate water, sanitation, and hygienetotals 1.7 million deaths and the loss of at least 50 million healthy life years. Some 1.1 billion people lack access to safe drinking water and 2.6 billion lack access to basic sanitation. Investments in drinking water supply and sanitation show a close correspondence with improvement in human health and economic productivity. Each person needs only 20 to 50 liters of water free of harmful contaminants each day for drinking and personal hygiene to survive, yet there remain substantial challenges to providing this basic service to large segments of the human population. Half of the urban population in Africa, Asia, and Latin America and the Caribbean suffers from one or more diseases associated with inadequate water and sanitation.

The state of freshwater resources is inadequately monitored, hindering the development of indicators needed by decision-makers to assess progress toward national and international development commitments.Substantial deterioration of hydrographic networks is occurring throughout the world, increasing the difficulty of making an accurate assessment of global freshwater resources. The same is true for groundwater monitoring, standard water quality monitoring, and freshwater biological indicators. New techniques make it possible to identify literally thousands of chemicals, including long-lived synthetic pharmaceuticals, in freshwater resources. But universal application of these techniques is lacking, and there are no systematic epidemiological studies to understand their impact on long-term human well-being.

Freshwater as a Provisioning Service

This chapter documents a growing dependence of human populations on these services, which has resulted in a variety of activities aimed at stabilizing and delivering water supplies. So effective has been the ability of water management to influence the state of this resource, in terms of both its physical availability and chemical character, that anthropogenic signatures are now evident across the global water cycle. Much of this influence is negative due to overuse and poor management. The capacity of ecosystems to sustain freshwater provisioning services is thus strongly compromised throughout much of the world and may continue to remain so if historic patterns of managed use persist.

Because the water cycle plays so many roles in the climate, chemistry, and biology of Earth, it is difficult to define it as a distinctly supporting, regulating, or provisioning service. Precipitation falling as rain or snow is the ultimate source of water supporting ecosystems. Ecosystems, in turn, control the character of renewable freshwater resources for human well-being by regulating how precipitation is partitioned into evaporative, recharge, and runoff processes. Together with energy and nutrients, water is arguably the centerpiece for the delivery of ecosystem services to humankind.

While ecosystems are strongly dependent on the water cycle for their very existence, at the same time these systems represent domains over which precipitation is processed and transferred back to the atmosphere as "green water" (through evapotranspiration drawn from soils and plant canopies in natural ecosystems and rain-fed agriculture). The remainder runs off as "blue water" which constitutes the renewable water supply that can pass to downstream users—both aquatic ecosystems and humans such as farmers who irrigate. These water flows can be tabulated across ecosystems to identify areas that are critical to human well-being as well as those that require particular attention in designing strategies for environmental protection.

Setting the Stage

Prior to the twentieth century, global demand for freshwater was small compared with natural flows in the hydrologic cycle. With population growth, industrialization, and the expansion of irrigated agriculture, however, demand for all water-related goods and services has increased dramatically, putting the ecosystems that sustain this service, as well as the humans who depend on it, at risk. While demand increases, supplies of clean water are diminishing due to mounting pollution of inland waterways and aquifers. Increasing water use and depletion of fossil groundwater adds to the problem. These trends are leading to an escalating competition over water in both rural and urban areas. Particularly important will be the challenge of simultaneously meeting the food demands of a growing human population and expectations for an improved standard of living that require clean water to support domestic and industrial uses.

Meeting even the most basic of needs for safe drinking water and sanitation continues to be an international development priority. Some 1.1 billion people lack access to clean water supplies and more than 2.6 billion lack access to basic sanitation. Reducing these numbers is a key development priority. By adopting the initial targets of the Millennium Development Goals, governments around the world have made a commitment to reduce by half the proportion of people lacking access to clean water supply and basic sanitation between 1990 and 2015.

The ministerial declaration from the 2nd World Water Forum in The Hague in 2000 captured the essence of the goals and challenges faced, including articulation of the importance of ecosystems in sustaining freshwater services. Water continues to rise in importance in major policy circles, with 2003 declared the International Year of Freshwater, release of the first World Water Development Report by a collaboration of 24 U.N. agencies through the World Water Assessment Programme, and proclamation by the UN General Assembly of the International Decade of Action "Water for Life" in 2005–15.

Societies have benefited enormously through their use of freshwater. However, due to the central role of water in the Earth system, the effects of modern water use often reverberate throughout the water cycle. Key examples of human-induced changes include alteration of the natural flow regimes in rivers and waterways, fragmentation and loss of aquatic habitat, species extinction, water pollution, depletion of groundwater aquifers, and "dead zones" (aquatic systems deprived of oxygen) found in many inland and coastal waters. Thus, trade-offs have been made—both explicitly and inadvertently—between human and natural system requirements for freshwater services.

The challenge for the twenty-first century will be to manage freshwater to balance the needs of both people and ecosystems, so that ecosystems can continue to provide other services essential for human well-being.

Monitoring the continental water cycle in a timely manner at the global scale using traditional discharge gauging stations— the mainstay of water resource assessment—continues to challenge the water sciences. Data collection is now highly project-oriented, yielding often poorly integrated time series of short duration, restricted spatial coverage, and limited availability. In addition, there has been a legal assault on the open access to basic hydrometeorologi-cal data sets, aided in large measure by commercialization and fears surrounding piracy of intellectual property. Delays in data reduction and release (up to several years in some places) are also prevalent.

Based on available global archives at the WMO Global Runoff Data Center, to which member states contribute voluntarily, there was arguably a better knowledge of the state of renewable surface water supplies in 1980 than today. Such statements apply to many parts of the world, including otherwise well monitored countries like the United States and Canada, though most marked declines are in the developing world. Our understanding of groundwater resources is even more limited, since well-log, groundwater discharge/ recharge, and aquifer property data for global applications are only beginning to be synthesized. Information on water use and operation of infrastructure has never been assembled for global analysis.

While remote sensing and models of the water cycle can be used to fill some data gaps, these approaches themselves produce a range of outputs arising from differences in their input data streams and detailed calculation procedures. Without a sustained international commitment to baseline monitoring, global water assessments will be difficult to make and fraught with uncertainty.

Trends in the Provision of Freshwater

While it is true that there is an abundance of water across blue planet Earth, only a small portion of it exists as freshwater, and even a smaller fraction is accessible to humans. Nearly all water on Earth is contained in the oceans, leaving only 2.5% as fresable. Of this small percentage, nearly three quarters is frozen, and most of the remainder is present as soil moisture or lies deep in the ground. The principal sources of freshwater that are available to society reside in lakes, rivers, wetlands, and shallow groundwater aquifers—all of which make up but a tiny fraction (tenths of 1%) of all water on Earth. This amount is regularly renewed by rainfall and snowfall and is therefore available on a sustainable basis.

Global averages fail to portray a complete picture of the world's water resource base, however. The basic climatology of the planet dictates that freshwater will be distributed unevenly around the globe, with abundant supplies across zones like the wet tropics and absolute water scarcity across the desert belts and in the rain shadow of mountains. For this assessment, both locally available runoff and water transported though river networks is considered. River corridor flows convey essential water resources to those living on the banks of large rivers, such as along the lower Nile.

The supply of freshwater is conditioned by several additional factors, which amplify the patterns of abundance and scarcity. These factors include the distribution of humans relative to the supply of water (that is, access to water), patterns of demand, presence of water engineering to stabilize flows, seasonal and interan-nual climate variations, and water quality.

Available Water Supplies for Humans

Estimates of global water supply are imprecise and complicated by several factors, including differences in data and methodologies used, loss of hydrographic monitoring capacity, alternative time frames considered, and distortions from land cover, climate, and hydraulic engineering that are increasingly a part of the water cycle. The renewable resource base expressed as long-term mean runoff has been estimated to fall between 33,500 and 47,000 cubic kilometers per year. Within-year variations also define the basic nature of water supply. At the continental scale, maximum-to-minimum runoff ratios vary between 2:1 and 10:1, with individual rivers experiencing ratios far higher, such as in snowmelt-dominated basins or episodically flooded arid and semiarid river systems. These variations necessitate flow stabilization through hydraulic engineering for either protection (for example, from floods) or seasonal supply augmentation (for example, for dry-season agriculture or hydroelectricity).

Water supply can also be assessed from the standpoint of societal access to renewable runoff and river flow, from which humans can secure provisioning services. By one estimate, one third of global renewable water supply is accessible to humans, when taking into account both its physical proximity to population and its variation over time, such as when flood waves pass uncaptured on their way to the ocean.

Groundwater plays an important role in water supply. It has been estimated that between 1.5 billion and 3 billion people depend on groundwater supplies for drinking. It also serves as the source water for 40% of self-supplied industrial uses and 20% of irrigation. For certain countries this dependency is even greater; for example, Saudi Arabia meets nearly 100% of its irrigation requirements through groundwater. Two important classes of groundwater can be identified. The first is renewable groundwater resources, closely linked to the cycling of freshwater, through which the ground is periodically replenished when sufficient precipitation is available to recharge soils or when floodplains become inundated. The second, fossil groundwater, is

The ranges reported here are from three global-scale water resource models, two of which were used directly in the MA: University of New Hampshire for the Condition and Trends Working Group assessment and Kassel University used in the Scenarios Working Group. A third model from the University of Tokyo and Global Soil Wetness Project was also compared.

The global-scale correspondence for total supply, withdrawals, water crowding, and demand-to-supply ratio is high, but masks continental-scale differences. Such disparities can be large, as for water supply in Latin

America, where large remote tropical river systems have proved difficult to monitor systematically. Substantial differences at the continental scale are noted for population living under severe water scarcity (use-to-supply >40%). The order-of-magnitude range apparent for sub-Saharan Africa can be linked in part to the distribution of sharp climatic gradients that are difficult to analyze geographically. The result is also a function of the assumptions made regarding access to water. Because of such uncertainties, the current state-of-the-art in global models put 1-2 billion people at risk worldwide arising from high levels of water use. The MA models predict a much smaller range, from 2.0-2.1 billion.

Large uncertainties surround current estimates of water consumption by the largest user of water, agriculture. Recent estimates vary from 900 up to 2000 cubic kilometers per year. A value of 1200 cubic kilometers per year is reported in this assessment.

Table 1. Major Storages Associated with the Contemporary Global Water System

Type	*Volume (percent)*	*Fraction of Freshwater (percent)*	*Fraction of Total Volume (thous. cu. km.)*
World ocean	1,338,000	96.5	-
Groundwaters	23,400	1.7	-
-Fresh	10,530	0.76	30.1
Soil moisture	16.5	0.001	0.05
Glaciers/permanent ice	24,100	1.74	68.7
Ice in permafrost	300	0.022	0.86
Lakes (fresh)	91	0.007	0.26
Wetlands	11.5	0.0008	0.03
Rivers	2.12	0.0002	0.006
Biological water	1.12	0.0001	0.003
Atmosphere	12.9	0.001	0.04
Total hydrosphere	1,386,000	100	-
Total freshwater	*35,029*	*2.53*	*100*

typically locked in deep aquifers that often have little if any long-term net recharge. Whenever this is extracted, it is functionally "mined," a particularly acute problem in arid regions, where replenishment times can be on the order of thousands of years.

Establishing the contribution of groundwater to the global supply of freshwater inserts a substantial element of uncertainty into the overall assessment. Problems of poor data harmonization, incomplete and fragmentary inventories, and methodological difficulties are well documented. As a result, there is large uncertainty in estimates of fresh groundwater resources, ranging from 7 million to 23 million cubic kilometers. While abundant, their use can be severely restricted by pollution or by the cost of extracting water from aquifers, which rises progressively in the face of extraction rates exceeding recharge.

Another important water supply is represented by the widespread construction of artificial impoundments that stabilize river flow. Today, approximately 45,000 large dams (>15 meters high or between 5 and 15 meters high and a reservoir volume of more than 3 million cubic meters) and possibly 800,000 smaller dams have been built for municipal, industrial, hydropower, agricultural, and recreational water supply and for flood control. Recent estimates place the volume of water trapped behind documented dams at 6,000– 7,000 cubic kilometers. In drainage basins regulated by large reservoirs (>0.5 cubic kilometers) alone, one third of the mean annual flow of 20,000 cubic kilometers is stored. Assuming seasonal six-month low flows constitute roughly 40% of annual discharge, this impounded water represents a global potential to carry over an entire year's minimum flows.

Desalinization constitutes a renewable water supply using distillation and membrane techniques to withdraw salt from otherwise unusable water. While the technology continues to improve, desalinization remains the most costly means of supplying freshwater and is highly energy-intensive. Costs range between $1 and $4 per cubic meter, placing it well above the most expensive traditional sources. Despite this, in 2002 there were over 10,000 desalinization plants in 120 countries supplying more than 5 cubic kilometers per year, with a global market of $35 billion per year. Collectively, these plants provide for much less than 1% of global freshwater use.

More than 70% of global installed desalinization capacity is in the oil-rich states of the Middle East and North Africa. While its use may be difficult to justify for high-water-consumptive activities like irrigation, investments in desalinization technologies are likely to improve efficiency and bring down costs, creating a potentially important source at least for domestic drinking water, and the annual supply of desalinized water could double in 15 years. The unresolved issue of adequately managing brine waste from the desalinization process to protect nearby coastal ecosystems requires special attention.

Finally, rainwater harvesting through traditional methods or modern technology is another way in which humans augment freshwater supply. Rainwater harvesting can directly increase the soil water content or be stored for later application as supplemental irrigation during dry periods. This is particularly important in places like India, which relies heavily on a short period of intense rainfall. The groundwater authorities in India, for instance, have made it mandatory for multistoried buildings in New Delhi and several other states to have a rooftop rainwater harvesting system. Rainwater harvesting can also be an appropriate technology for maintaining groundwater base flow and reducing flood peaks.

Total Flows of Freshwater

Ecosystems vary greatly in their exposure to precipitation and hence as source areas for renewable runoff that emerges as part of the hydrologic cycle. The proportional contribution of each ecosystem to global runoff is generally equivalent to the fraction of precipitation to which it is exposed. Forests therefore are associated with slightly more than half of global precipitation and yield about half of global runoff, while mountains represent one quarter of both global precipitation and runoff. Cultivated and island systems are the next most important source areas, each constituting about 15% of global runoff.

All other systems contribute 10% or less. Paradoxically, dryland ecosystems, due to their large aerial extent, receive a nearly identical fraction of global precipitation as mountains do, yet because of substantial losses from the system due to evapotranspiration, they are a relatively minor contributor to global renewable water supply (<10%). Urban systems, because of their restricted extent (<<1% of land area), receive only 0.2% of global precipitation and provide the same very minor proportion of global runoff.

From a regional perspective, Latin America is most water-rich, with about one third of global runoff. Asia is next, with one quarter of global runoff, followed by OECD (20%), and sub-Saharan Africa and the former Soviet Union, each with 10%. The Middle East and North Africa is clearly driest and most water-limited, accounting for only 1% of global runoff.

Freshwater Flows Accessible to Humans

Ecosystems constitute the ultimate source areas for freshwater provisioning services. The accessibility of renewable water supply can be estimated through an index measuring the proportion of total annual renewable runoff generated locally that eventually flows through river corridors and encounters downstream human populations. The importance of upstream ecosystems as source areas for freshwater supply is demonstrated. Cultivated, coastal, and urban systems, with sizable fractions of the global population, have from 90% to 100% of their renewable runoff accessible. Drylands also show high accessibility, likely reflecting the propensity of humans to settle near scarce freshwater resources. Mountains, forests, and inland waters each show 70–80% of total runoff as accessible to downstream populations. The exception is polar systems, which yield less than 20% of total runoff as accessible, reflecting their remote and generally uninhabited environment.

Populations served by accessible runoff emerging from individual ecosystems are typically in the billions. Cultivated systems, forests, inland waters, and mountains each serve at least 4 billion people. Four fifths of the world lives downstream of runoff from cultivated lands, followed by a nearly identical fraction downstream from forests. Inland waters and mountains provide water to two thirds of global population and drylands to one third. Remote islands and polar systems serve the fewest people. Runoff from urban systems, nearly all generated in close proximity to densely settled areas, serves nearly three quarters of the world's population.

The large fractions of total runoff expressed as accessible runoff indicate that, by and large, human society has positioned itself into areas with identifiable local sustainable water supplies or river corridor flows. A geographic distribution of human settlement thus is linked to the availability of freshwater. Mountains serve 3 times, forests 4 times, and inland waters 12 times as many people downstream through river corridors as they do through locally derived runoff. Urban areas nearly double the total service when tabulating downstream populations. Remaining ecosystems show more-limited importance in transferring precipitation as accessible runoff to downstream populations. For drylands, this is due to a lack of substantial quantities of runoff, while for coastal or island systems it is a consequence of short flow pathways to the ocean. Each of these systems still supplies 15–30% of global population with renewable and accessible runoff.

From a regional perspective, Latin America and Asia constitute the largest proportion (together nearly 60%) of global accessible runoff. And while the OECD, sub-Saharan Africa, and the former Soviet Union generate a large portion of the global runoff, substantial quantities are remote and inaccessible particularly in the former Soviet states. The Middle East and North Africa generates less than 1% of renewable accessible runoff.

Overall, the global fraction of total annual runoff that is accessible to humans is 75%, with slightly more than 80% of world population (4.9 billion people) being served by these renewable and accessible water flows. However, while providing an estimate of long-term water supply, these figures overstate the effective availability of freshwater. Given that approximately 30% of annual runoff is uncaptured flood flow, the world's population has its access reduced from 75% to 53% of total runoff.

Globally, renewable freshwater services reflect the geographic distributions of both water supply and human populations. Four out of every five people live downstream of and are served by renewable freshwater services. Thus, while the human population is generally well organized with respect to the availability of freshwater, 20% of humanity remains without any appreciable quantities of sustainable supply or must gain access to such resources through costly interbasin transfers from more water-rich areas. These people are highly reliant on unsustainable water resources. For those with access to renewable supplies, a total of 65% of the world's population is served by the 50% of total annual renewable runoff that is positioned in dry to moderately wet conditions, with concomitant pressure on that resource base. Only 15% live with relative water abundance—that is, in conjunction with the remaining 50% of total runoff (represented by the high runoff-producing regions shown in the upper part of the curve. If uncaptured flood flow is incorporated into these calculations, for the 80% of world population who reside in the lower half of the water availability spectrum with no appreciable renewable freshwater flows), the effective supply is reduced from 50% to 35% of total runoff.

Water Use

Over the last few centuries, global water use has shown roughly an exponential growth and been linked closely to both population growth and economic development. There was a fifteenfold increase in global water withdrawals between 1800 and 1980, when population increased by a factor of four. Since the 1900s, the overall increase has been sixfold. Global consumptive water losses, primarily from evapotranspiration through irrigation, increased thirteenfold during this same period. A major, recent feature of human water use is the reduction in per capita use rates, dropping as of around 1980 from about 700 to 600 cubic meters per year, though the aggregate global withdrawal continues to increase.

While the general features of a historical rise in freshwater demands are clear, there are substantial uncertainties surrounding water use estimates, reflecting the current state of knowledge, assumptions (or lack thereof) on potential efficiency changes and reuse potential, number of years projected into the future, and interactions with market forces. The summary statistics from three global tabulations provided earlier, demonstrate the current degree of uncertainty.

Global water withdrawals today total about 3,600 cubic kilometers per year, with a wide range of use over individual continents. The largest user is Asia, accounting for nearly half of the world total, with OECD next, using about one third. The remaining continents each represent less than 10% of global use. Water use today is dominated by agricultural withdrawals (70% of all use), followed by industrial and then domestic applications. Withdrawals in agriculture are fundamentally defined by irrigation. In Asia, the Middle East and North Africa, and sub-Saharan Africa, agriculture accounts for 85–90% of all withdrawals. Driven by irrigation demand, overall withdrawals across MENA constitute 120% of renewable accessible supplies, meaning that this region relies on nonrenewable supplies for food production. Agricultural water use in the former Soviet Union and the OECD is proportionally much lower, reflecting the water needs of other sectors in these industrial economies. In contrast, industrial water use is only 4% in sub-Saharan Africa, reflecting a low level of economic development.

Water lost from groundwater and surface water sources to the atmosphere through net evaporation (such as from irrigation, cooling towers, or reservoirs) is termed water consumption or irretrievable losses, which today represent a substantial fraction of water use. Contemporary irretrievable losses through irrigation, computed as the evapotranspiration component of agricultural withdrawals, are assessed here. Irretrievable losses from irrigation represent one third of all water use globally. The efficiency computed for irrigated agriculture (the ratio of water withdrawn to water consumed or lost through evapotranspiration on irrigated cropland) is on average 50% globally and varies from 25% (in Latin America) to 60% (in Asia). Additional losses from evaporation from reservoirs, irrigation ditches, and so on are difficult to estimate accurately but could total over 500 cubic kilometers per year, thus indicating the conservative nature of the consumption estimates.

Non-sustainable water use could be a substantial component of total withdrawals. Earlier work based on documentary evidence showed approximately 200 cubic kilometers per year of global aquifer overdraft, though the estimate is regarded as highly uncertain. This assessment of water supply and use (based on Vorosmarty et al. 2000, 2005; Fekete et al. 2002) using a geospatial framework (about 50-kilometer resolution) enables calculations to be made of the degree to which water withdrawal exceeds locally accessible supplies— in other words, non-sustainable water use (U_{an}). Worldwide, non-sustainable withdrawals can be computed using two endpoints: crop evaporative demands or water use statistics, which include both consumption and transport losses, some unknown fraction of which reenters the surface-groundwater system for potential reuse. These endpoints give a calculated non-sustainable use of about 400–800 cubic kilometers per year. In terms of total freshwater withdrawals, 10–25% could represent nonrenewable use. When the earlier estimate of 200 cubic kilometers per year is also included, a large degree of uncertainty results, and from 5% to 25% of freshwater withdrawals could represent nonrenewable use.

Nevertheless, each of these estimates reflects a high dependence on existing water services, especially in areas where induced, chronic water stress necessitates costly water engineering remedies, groundwater depletion, or curtailment of water-using activities. Each continent shows a heavy reliance on such nonrenewable extraction, ranging up to one third of total use based on the high estimates. Asia and MENA show the greatest level of such dependence; OECD, the least. In MENA, 30% of all water use is from non-sustainable sources, and this use is equivalent to over one third of accessible renewable supplies.

It shows the contemporary geography of such non-sustainable use and demonstrates the much larger impacts that arise at subcontocally. The spatial pattern of overuse is broadly consistent with previously reported regions of use exceeding supply, major water transfer schemes, or groundwater overdraft: Australia, western Asia, northern China, India, North Africa, Pakistan, Spain, Turkey, and the western United States.

Non-sustainable use expressed as a proportion of irrigated agricultural withdrawals shows an even higher degree of dependency on nonrenewable supplies. Globally, about 15–35% of irrigation withdrawals are computed to be non-sustainable. Individual continental areas show percentages ranging from less than 10% to 40%, as in the case of Asia. Such high rates indicate an increasing degree of food insecurity. Given projections showing no major expansion in global cropland area, increasing pressure will be placed on irrigated cropland, which today provides nearly 40% of crop production. By its very nature, this water use cannot persist indefinitely, and many regions of the world have well-documented cases of aquifer depletion and abandonment of irrigation, adding constraints to irrigated crop production arising from rising development costs, soil salinization, and competition for water required by sensitive ecosystems and commercial fisheries.

Notion of Water Scarcity

The assessment thus far has shown a growing dependence of human society on accessible freshwater resources. To assess the state of these provisioning services more comprehensively, the supply of renewable water must be placed into the context of interactions with people and their use of water. A set of relative measures can be used in this regard.

One measure of dependence on freshwater is the population served per million cubic meters per year of accessible runoff (renewable supply). This is known as the "water crowding" index, with levels on the order of 600–1,000 people per million cubic meters per year (that is, 1,000–1,700 cubic meters per year supply per person) showing water stress, and above 1,000 people (that is, less than 1000 cubic meters per year per person) indicating extreme water scarcity. Another measure is the relative water use or water stress index, expressed as the ratio of water withdrawals to supply.

Worldwide, a substantial quantity of renewable freshwater supply—nearly 30,000 cubic kilometers per year—is accessible to humans. Thus contemporary use represents slightly more than 10% of annual supply. However, there is a substantial range in the share of accessible runoff used by humans across different continents as well as a rapidly changing picture over the last few decades. Time series of use indicate increasing pressures on the freshwater resource base.

Between 1960 and 2000, world water use doubled from about 1,800 to 3,600 cubic kilometers per year, a rate of about 17% per decade, with a slower (10%) increase projected to 2010. Individual continents show increases over the 1960– 2000 timeframe from 15% up to 32% per decade. MENA has historically shown a great dependence on its freshwater supply, using well over half as early as 1960 and exceeding all renewable supplies shortly after 1980. Today its withdrawals represent 120% of accessible sustainable supply, and these are projected to rise to >130% by 2010. Asia, the former Soviet Union, and OECD countries show intermediate levels of use relative to supply over this period. In sub-Saharan Africa, substantial contributions of freshwater from river basins in the wet tropics coupled with rela-tively poor water delivery infrastructure and restricted development mean that only 2% of renewable supply is tapped. In water-rich Latin America, relative use rates also remain low, at less than 5%.

The contemporary water crowding index is modest in almost all regions. Only MENA shows a value reflective of its well-known position as a highly water-scarce region. Over the last four decades there has been a sustained and substantial increase in the water crowding index with respect to accessible runoff, reflecting directly the impact of population growth. Worldwide, the number of people served per unit of supply has doubled during this period, at an average rate of 20% per decade. Several regions show even greater rates of increase—a tripling for MENA and sub-Saharan Africa and a more than doubling for Asia and Latin America. Globally, an additional 13% crowding in renewable supply is predicted between 2000 and 2010, with greatest regional increases expected in sub-Saharan Africa (30%) and MENA (20%). A slight slowing in rate of increase is noted globally, with near stability in the index for OECD and the former Soviet states.

Several cautionary notes are needed in interpreting these trends. The statistics are based on mean annual flows and access computed for 100% of individual continental and global populations. In the context of the 50% of continental runoff generated in dry to moderately wet climate zones (19,800 cubic kilometers per year) that serves the majority of global population, contemporary use represents nearly 20% of the mean annual supply. When seasonal variations in runoff are considered (reducing supplies to 13,900 cubic kilometers per year), withdrawals exceed 25% of the renewable resource. In addition, if dedicated instream uses of about 2,000 cubic kilometers per year for navigation, waste processing, and habitat management are considered, humans then use and regulate 40% or more of renewable accessible supplies.Further, the crowding index does not take into account different countries' abilities to deal with water shortages. For example, high-income countries that are water-scarce may be able to cope to some degree with water shortages by investing in desalination or reclaimed wastewater. The study also discounts the use of fossil water sources because such use is unsustainable in the long term.

In addition, while the global numbers are well below the extreme scarcity threshold of 1,000 people per million cubic meters per year of renewable supply, they mask important local and regional differences and thus understate the true degree of stress. Prior assessments show that as of 1995 some 41% of the world's population, or 2.3 billion people, were living in river basins under water stress, with some 1.7 billion of these people residing in river basins under conditions of extreme water scarcity. From a river basin perspective, the Volta, Nile, Tigris and Euphrates, Narmada, and Colorado in the United States will show ongoing pressure through 2025. Another 29 basins will descend further into scarcity by 2025, including the Jubba, Goda-vari, Indus, Tapti, Syr Darya, Orange, Limpopo, Yellow, Seine, Balsas, and Rio Grande. Indicators based on mean annual conditions also mask important supply limits imposed by seasonal and inter-annual variability. For example, in India most of the annual water supply is generated as a result of the monsoons, which in many cases means both flooding downstream as well as seasonal drought.

Another measure of adequacy of the freshwater supply is the mean use-to-supply ratio. A set of thresholds for water stress was given by the United Nations in a recent global analysis that used this ratio based on mean annual conditions : low (<10%), moderate (10–20%), medium/high (20–40%), and high (>40%). Using this classification and a grid-based approach necessary to capture the high degree of spatial heterogeneity, the contemporary global-scale ratio is from low-to-moderate, as seen, although entire continents are under a moderate (Asia, former Soviet Union, and OECD) to high (MENA) state of scarcity. This is in stark contrast to the situation in 1960, when uniformly low levels of scarcity were noted (with the exception of MENA). Globally, it has been shown that 2.5 billion people suffer from at least moderate levels of chronic water stress and from 1–2 billion people suffer high levels of scarcity even when tabulations are made conservatively on total renewable supplies. Calculating the population at risk through a ratio based osn accessible supplies would increase the overall exposure to stress.

Water scarcity as a globally significant problem is a relatively recent phenomenon, evolving only over the last four decades. Rates of increase in the relative use ratio from 1960 to the present averaged about 20% per decade globally, with values from 15% to more than 30% for individual regions. A slowing in the rate of increase in use is projected between 2000 and 2010, to 10% per decade globally. With anticipated population growth, economic development, and urbanization, a further increase in the relative use ratio for some continents is likely to remain high.

Environmental Flows for Ecosystems

In light of the expanding use of freshwater by humans and several indicators of growing water stress, an important issue emerges with respect to the sustainability of water provisioning services— that is, being able to continue providing water for human use while also meeting the water requirements of aquatic ecosystems so as to maintain their capacity to provide other services. "Environmental flows" refers to the water considered sufficient for protecting the structure and function of an ecosystem and its dependent species. These flow requirements are defined by both the long-term availability of water and its variability and are established through environmental, social, and economic assessment.

Determining how much water can be allocated to human uses or distorted through flow stabilization (such as dam construction) without loss of ecosystem integrity is central to an understanding of how freshwater ecosystems support human well-being through the range of provisioning, supporting and regulating services. Assessment of water availability, water use, and water stress at the global scale has been the subject of on-going research. However, water requirements of aquatic ecosystems are only now being estimated globally and considered explicitly in these assessments. Flow requirements can range globally from 20% up to 80% of mean annual flow, depending on the river type, its species composition, and the river health condition objectives sought (for instance, pristine, moderate modification from natural conditions, minimum flows), indicating the high degree of potential conflict with river regulation and human uses should the environment be preserved.

If human systems are viewed as being embedded within natural systems, human water use can expand to a "sustainability boundary" beyond which a substantial degradation of ecosystem services results. Determining the location of the sustainability boundary is critical to successful management and rests on clearly defining what constitutes a degraded ecosystem. Environmental flows should consider both the quantity and timing of flow to maintain "naturally variable flow regimes", whereby seasonal flow patterns are maintained with the aim of retaining the benefits provided by low and high flows. Naturally low flows, for example, help exclude invasive species while high flows, especially floods, shape channels and allow the delivery of nutrients, sediments, seeds, and aquatic animals to seasonally inundated floodplains. High flows may also provide suitable migration and spawning cues for fish.

Global Trends in Water Diversion and Flow Distortion

While global trends in altered water regime are difficult to assemble with certainty due to incomplete information, they reflect an overall increase in regulation of the world's inland river systems. Water withdrawals show a doubling between 1960 and 2000, by which time irretrievable losses from irrigation alone totaled 34% of all global use.

One third of all rivers for which contemporary and pre-disturbed discharges could be compared in a compendium showed substantial declines in discharges to the ocean. Long-term trend analysis (more than 25 years) of 145 major world rivers indicated more than one fifth with declines in discharge. From 1960 to 2000 there was a near quadrupling of reservoir storage capacity and more than a doubling of installed hydroelectric capacity. Worldwide, large artificial impoundments (storing each 0.5 cubic kilometers or more) now hold two to three months of runoff, capable of significant hydrograph distortion, with several major basins showing storage potentials of greater than a year's runoff. Much of this regulation occurred over the last 40 years.

Through consumptive use and interbasin transfers, several of the world's largest rivers (Nile, Yellow, Colorado) have been transformed into highly stabilized and in some cases seasonally nondischarging river channels. In the case of the Yellow River, improved water management since 2000 has helped to restore flows.

Recent History of Governance and Management for Environmental Flows

Over the last decade, policy solutions to developing environmental flows have taken several forms, depending on social and historical context, degree of scientific knowledge, water infrastructure, and local ecosystem conditions.

These approaches include managing the quantity and temporal pattern of water withdrawals or releases, developing water markets, and preemptively managing land use to protect watersheds.

Water allocation for environmental flows to sustain functioning freshwater ecosystems is practiced in parts of Australia, Europe, New Zealand, North America, and South Africa. However, there appears to be very little consideration of this matter anywhere in Asia, despite aggressive water extraction from many rivers during the dry season across the continent. But there is cause for cautious optimism. The calculation, adoption, and implementation of environmental flows are under consideration in other parts of the world. In addition, more than 2,000 river, lake, and floodplain restoration projects in at least 20 countries, particularly in Europe but also in Africa and Asia, are being carried out. Some key examples include the restoration of the Diawling delta in Mauritania, the Waza Logone flood-plain in Cameroon, the Danube and Rhine Rivers, and the South Florida Everglades—one of the largest ecosystem restoration projects ever attempted.

The shift toward management for natural flow regimes is also reflected by parallel shifts in public policy from laws favoring private interests and prior appropriations (as in much of the American West) to protecting water rights and environmental flows as part of the "public trust." In 1998, South Africa passed landmark legislation to aid decision-making on all or part of any significant water resource. One of the most progressive aspects of this act was establishment of a Reserve to support both essential human needs (water for drinking, food preparation, personal hygiene) and aquatic ecosystem integrity. Notably, this two-part Reserve—with human and environmental components—takes priority over other uses such as irrigation and industrial withdrawal. In Burkina Faso, a new water framework law (Loi d'Orientation sur L'eau), adopted in 2001, establishes the legal and institutional framework for promoting integrated basin management, equitable access, water for nature, and international cooperation. The legislation recognizes that "infrastructures which are built on a water course must maintain a minimal flow that guaranties aquatic life".

For many highly regulated river systems in North America (e.g., Colorado, Columbia, Missouri, Savannah), recent changes in dam operations and adaptive management plans are now fostering conditions that improve fish habitat, river-floodplain connectivity, and estuarine ecosystems, often at the cost of hydroelectric generation or navigability to barges (Postel and Richter 2003; Richter et al. in prep.). In addition, the decommissioning and removal of some dams has begun in the United States. In Australia, water allocation reforms have led to limits on future withdrawal (that is, a "water cap") in the Murray-Darling River basin, subsequent development of a water market where allocations are traded, and creation of incentives to increase water productivity and efficiency. Similarly, water markets developed in Mexico, Chile, and some western states in the United States have been used to secure flows for ecosystems.

Watershed management strategies that integrate ecological principles have been used to prevent water supply crises from developing. An often-cited example is the New York City water supply management strategy, which includes protection of riparian habitat in the nearby source area of the Catskills Mountains, thus eliminating the need to construct a water filtration plant at an estimated cost of $6 billion. The ~400,000-hectare Pinelands National Reserve in nearby New Jersey is regulated under a Comprehensive Management Plan developed at the local, state, and federal level in 1978–79. The plan permits a wide spectrum of land use development categories, ranging from intensive development to full protection, and it successfully redirected human activities to areas deemed appropriate while protecting a large core area, which is ecologically sensitive, drought-prone, and nutrient-poor and which harbors a unique community of wildlife with a large number of endemic species. The benefits of maintaining high water quality are recognized outside the reserve through the delivery of relatively high-quality freshwater to an estimated 9 million people in New York City for less than if a water filtration plant were built. In addition, water discharged into Delaware Bay helps to support populations of anadromous fish and spawning horseshoe crabs, which in turn support large numbers of migrating shorebirds and local industries.

Water Quality

Summarizing patterns and trends in water quality, particularly at a global scale, encompasses an array of challenges that include basic definitional problems, a lack of worldwide monitoring capacity, and an inherent complexity in the chemistry of both natural and anthropogenic pollutants. From a management perspective, water quality is defined by its desired end use. Water for recreation, fishing, drinking, and habitat for aquatic organisms thus require higher levels of purity, whereas for hydropower, quality standards are much less important. For this reason, water quality takes on a broad definition as the "physical, chemical, and biological characteristics of water necessary to sustain desired water uses".

Natural water chemistry is inherently highly variable over space and time, and aquatic biota are adapted to this variability. With added pressure from human activities, the biogeophysical state of inland waters plus their variability is altered, often to the detriment of aquatic species, thereby compromising the sustainability of aquatic ecosystems. Many chemical, physical, biological, and societal factors affect water quality: organic loading (such as sewage); pathogens, including viruses in waste streams from humans and domesticated animals; agricultural runoff and human wastes laden with nutrients (such as nitrates and phosphates) that give rise to eutrophication and oxygen stress in waterways; salinization from irrigation and water diversions; heavy metals; oil pollution; literally thousands of synthetic and persistent engineered chemicals, such as plastics and pesticides, medical drug residues, and hormone mimetics and their by-products; radioactive pollution; and even thermal pollution from industrial cooling and reservoir operations.

Furthermore, despite important improvements in analytical methodologies, the capacity to operationally monitor contemporary trends in water quality is even more limited than monitoring the physical quantity of water. In terms of the spatial coverage, frequency, and duration of monitoring, data currently available for global and regional-scale assessments are patchy at best, leading to oversimplified and sometimes misleading information.

Data abundance is generally associated with level of economic development: industrial countries show a higher level of data availability, while water quality in developing countries is less well monitored. Even when data from monitoring stations are available, they only provide a fragmented view of water quality issues for very local sections of rivers, necessitating potentially unreliable extrapolation to the rest of the basin. For this reason, water quality assessments or trajectories are usually river-or station-specific. Even for the best-represented regions of the globe, a coherent time series of data is available for only the last 30 years or less, constraining the ability to clearly quantify trends in water quality.

Data comparability problems are yet another constraint on the utility of water quality data. Standardized protocols, in terms of sampling frequency, spatial distribution of sampling networks, and chemical analyses, are still not in place to ensure the production of comparable data sets collected in disparate parts of the world. The monitoring of groundwater supplies is even more problematic ; because ground- water is hidden from view, many pollution and contamination problems that affect supplies have been more difficult to detect and have only recently been discovered.

These many factors make it difficult to estimate the impact of changing water quality on global water supply. The following sections provide an overview assessment of trends in water quality that have bearing on the capacity of the contemporary water cycle to provide provisioning services for freshwater and on the sus-tainability of inland water systems. Other assessments specifically target water quality issues over selected regional-to-continental domains.

General Trends in Water Quality

The state of inland water quality illustrates the long-term and complex nature of human interactions with their environment. The earliest changes attributable to humans likely occurred in tandem with land use change in small to medium-sized catchments some 5,000 or 6,000 years ago in the Middle East and Asia, where water and sediment budgets were substantially altered. Water also has been considered since ancient times to be the preferred medium for cleaning, transporting, and disposing of wastes—establishing a tradition that today has substantially transformed the physical, biological, and chemical properties of global runoff.

A set of syndromes depicting riverine changes arising from anthropogenic pressures has been proposed through which society transforms inland freshwaters from a pristine state fully controlled by the natural Earth system to a modern condition in which humans provide many of the predominant controls. In most of the densely populated areas of the world, river engineering, waste production, and other human impacts have significantly changed the water and material transfers through river systems to the extent that this now likely exceeds the influence of natural drivers. This is true today in many parts of the Americas, Africa, Australasia, and Europe.

The contrast between pristine and contemporary states can be dramatic and potentially global in scope. Changes to the global nitrogen cycle are emblematic of those in water quality more generally, through which high concentrations of people or major landscape disturbances (such as industrial agriculture) translate into a disruption of the basic character of natural water systems. In addition, modern changes often "reverberate" far downstream of the original point of origin. Compared with the preindustrial condition, loading of reactive nitrogen to the landmass has doubled from 111 million to 223 million tons per year or possibly 268 million tons. Model results show these accelerated loadings transformed into elevated freshwater transports through inland waterways to the coastal zone, doubling pre-disturbance rates from 21 million to 40 million tons per year. North America, continental Europe, and South, East and Southeast Asia show the greatest change.

Riverine transport of dissolved inorganic nitrogen (immediate precursors to nutrient pollution, algal blooms, and eutrophica-tion) have increased substantially from about 2–3 million tons per year from the preindustrial level to 15 million tons today, with order-of-magnitude increases in drainage basins that are heavily populated or supporting extensive industrial agriculture. Rivers with high concentrations of inorganic nitrogen constitute a major global source for inorganic nitrogen, despite relatively modest contributions to aggregate water runoff. While it is noteworthy that aquatic ecosystems "cleanse" on average 80% of their global incident nitrogen loading, the intrinsic self-purification capacity of aquatic ecosystems varies widely and is not unlimited. As a result, sustained increases in loading from land-based activities are already reflected in the deterioration of water quality over much of the inhabited portions of the globe, they extend their impacts to major coastal receiving waters, and they are likely to continue well into the future.

While the stark contrast between pristine and contemporary states demonstrates the overall impact of anthropogenic influences on water quality, much of the contamination of freshwater has occurred over the last century. The main contamination problems 100 years ago were fecal and organic pollution from untreated human wastewater. Even though this type of pollution has decreased in the surface waters of many industrial countries over the last 20 years, it is still a problem in much of the developing world, especially in rapidly expanding cities.

In developing countries, sewage treatment is still not commonplace, with 85–95% of sewage discharged directly into rivers, lakes, and coastal areas, some of which are also used for water supply. Consequently, water-related diseases, such as cholera and amoebic dysentery, among others, claim millions of lives annually. In Europe, organic pollution and contamination by toxic metals are probably now less than the levels observed between the 1950s and1980s, due to improved environmental regulation. In the developing world, the riverine evolution is likely to be similar to that found in Europe, with a major lag corresponding to their different stages of industrialization, urbanization, and intensification of agriculture.

New pollution problems from agricultural and industrial sources have emerged in industrial and developing countries and have become one of the biggest challenges facing water resources in many parts of the world. In Western Europe and North America, on the one hand phosphorus contamination in waterways has been reduced considerably with the introduction of phosphate-free household detergents, investments in wastewater treatment plants, and to some degree modified agroecosys-tem management. On the other hand, residues of synthetic pharmaceuticals for humans and livestock are increasingly being discovered at low doses in rivers and lakes. There are indications that these residues can disturb the physiology of invertebrates, and it is still a matter of debate whether and, if so, to what degree these newly discovered pollutants may affect human physiology.

Water contamination by pesticides has grown rapidly since the 1970s. In a medium-sized river basin like the Seine, over 100 different types of active molecules from pesticides can be found. Even if the use of xenobiotic substances is increasingly being regulated in Western Europe and North America, bans—when they exist—occur generally two to three decades after the first commercial use of the products. For example, DDT, atrazine (a common pesticide), and PCBs were in use for a long time before they were banned in parts of the industrial world. In general these bans take longer to implement in the developing world, so these products are still commercialized and used in some countries.

In the United States, PCB and DDT records in estuarine sedimentary archives peaked in the 1970s and are now markedly decreasing. At the same time, persistent xe-nobiotics are widespread, with a recent study finding traces of at least one drug, endocrine-disrupting compound, insecticide, or other synthetic chemical in 80% of samples from 139 streams in 30 states of the United States. The persistence of these products in continental aquatic systems can be high, and their degradation products can be more toxic than the parent molecules. Because of the poor monitoring of the long-term effects of xenobiotics, the global and long-term implications of their use cannot be fully assessed.

Global Ranking of Water Quality Issues Based on Regional Assessment

A global water quality assessment, originally as part of the Dublin International Conference on Water and the Environment and in preparation for the Rio Summit is summarized here. The original report determined a global ranking of key water quality issues based on UN. Global Environmental Monitoring System data, the perceptions of local/regional scientists and managers, published reports and papers, and expert knowledge. Lakes, groundwater, and reservoir issues were considered, although as Siberia and northern Canada were not expressly covered in the 1991 report, these have been considered separately using the same approach. Eleven variables were considered and ranked, the scoring of which ultimately reflects the aggregate impact of human pressures, natural rates of self-purification, and pollution control measures.

The results show that pathogens and organic matter pollution (from sewage outfalls, for example) are the two most pressing global issues, reflecting the widespread lack of waste treatment. As water is often used and reused in a drainage basin context, a suite of attendant public health problems arises, thus directly affecting human well-being. At the other extreme, acidification is ranked #10 and fluoride pollution #11. The importance of the various issues varies between regions, however, and some of these globally low-ranked issues are particularly important in certain areas, such as acidification in Northern Europe, salinization in the Arabic peninsula, and fluoride in the Sahel and African Great Lakes. Fluo-ride and salinization issues are mostly due to natural conditions (rock types and climate), but mining-related salinization can also be found (for instance, in Western Europe). All other concerns directly arise through human influences.

Although these updated results correspond well to the state of water quality in the 1980–90s, since the 1990s the situation in most developing countries and countries in transition is likely worse in terms of overall water quality. In Eastern Europe, Central and South populated Americas, China, India, and populated Africa, it is probably worse for metals, pathogens, acidification, and organic matter, while for the same issues Western Europe, Japan, Australia, New Zealand, and North America have shown slight improvements. Nitrate is still generally increasing everywhere, as it has since the 1950s. In the former Soviet Union there has been a slight improvement in water quality due to the economic decline and associated decrease in industrial activities. Eastern Europe has also seen some improvements, such as those in the Danube and the Elbe basins. A few rivers, such as the Rhine, have seen a stabilization of nitrate loads after 1995.

Drivers of Change in the Provision of Freshwater

The drivers of change in the global water cycle and the system's capacity to generate freshwater provisioning services act on a variety of spatial and time scales. Throughout history, humans have pursued a very direct and growing role in shaping the character of inland water systems, often applied at local scales, but sometimes reflecting provincial or national policies on water. The collective significance of human influences on the hydrologic cycle may today be of global significance, but this has only recently begun to be articulated.

Humans today control and use a significant proportion of the runoff—from 40% to 50% —to which the vast majority has access. Given high numbers of people dependent on water provisioning services derived from ecosystems and the growing degree of water crowding, urbanization, and industrialization, the global water cycle is and will continue to be affected strongly by humans.

Water engineering to facilitate use by humans has fragmented aquatic habitats, interfered with migration patterns of economically important fisheries, polluted receiving waters, and compromised the capacity of inland water ecosystems to provide reliable, high-quality sources of water. Land cover changes have also altered the patterns of runoff and created sources of pollution, negatively affecting human health, aquatic ecosystems, and biodiversity. Due to a growing reliance on irrigated agriculture for domestic food production and international trade, freshwater services—in decline in many parts of the world through non-sustainable resource use practices—are directly linked to the global food security issue. Finally, natural climate variability and anticipated changes associated with greenhouse warming convey additional, major constraints on the provision of renewable freshwater services.

Population Growth and Development

Population growth is a major indirect driver of change in the provision of freshwater. Although freshwater supplies are renewed through a more or less stable global water cycle that produces precipitation in excess of evapotranspiration over the continents, the mean quantity of water supply available per capita is ever-decreasing due to population growth and expanding con-sumptive use. Human population doubled from 1960 until the present, and nearly 20 contemporary cities are home to 10 million people or more. Substantial flow stabilization and increased withdrawals have occurred across all regions, supporting an increase in the number of people sustained by the accessible, renewable water supply.

Continued growth in population will fuel increases in food production, which in the context of a stable cropland base will require greater diversions of freshwater for irrigation or considerably more efficient use of water supplies. The same applies to industry and municipalities, amplifying current pressures on the global water supply. Economic development, technology, and lifestyle changes (such as increasing meat consumption) further define the functional availability of water in the context of declining per capita supplies. Over the twentieth century, water withdrawals increased by a factor greater than six—more than twice the rate of population growth.

Managed Water Supplies

A broad array of water engineering schemes has enabled variability in the hydrologic cycle to be controlled and increasing amounts of water to be stored and withdrawn for human use. This technology refers to any sort of engineering used in the storage, management, and distribution of water, such as dams, canals, water transfers, irrigation ditches, levees, and so on. It also includes both traditional water harvesting techniques as well as modern production and treatment facilities like desalinization plants.

Global patterns of water management are not driven solely by investments in technology and large-scale engineering. Water is also managed through international trade, by way of the embodied or "virtual" water content of commodities exchanged. The agricultural sector, in particular, requires huge amounts of rainfall or irrigation water, much of which is lost to evapotranspiration, and in the case of irrigation there are also transit losses. Water input-to-crop output ratios, expressed on a weight-to-weight basis, vary from the hundreds to the thousands.

Role of Engineering on Water Supply

Dams and reservoirs

Humans have altered waterways around the world since historical times to harness more water for irrigation, industry, and domestic and recreational use. Dams have been a particularly significant driver of change, buffering against both spatial and temporal scarcity of water supplies and increasing the security of water and food supply over the past half-century. However, large engineering works that impound and divert freshwater have caused damage to key habitats and migratory routes of important commercial and subsistence fisheries, as well as serious societal disruptions, including public health problems and forced displacements.

Large dams are today the fundamental feature of water management across the globe. Approximately 45,000 large dams (>15 meters in height) and possibly 800,000 smaller dams are in place and an estimated $2 trillion has been invested in them over the last century. These facilities have served as important instruments for development, with 80% of the global expenditure of $32–46 billion per year focused on the developing world.

Major stabilization of global river runoff from major engineering works expanded greatly between 1950 and 1990. Currently the largest reservoirs—those with more than 0.5 cubic kilometers of storage capacity—intercept locally 40% of the water that flows off the continents and into oceans or inland seas. The volumetric storage behind all large dams represents from three to six times the standing stock of water held by natural river channels. In addition, large reservoir construction has doubled or tripled the residence time of river water—that is, the average time that a drop of water takes to reach the sea, with the mouths of several large rivers showing delays on the order of many months to years.

Such regulation has enormous impacts on the water cycle and hence aquatic habitats, suspended sediment, carbon fluxes, and waste processing. Large dams, in particular, have been a controversial component of the freshwater debate. While contributing to economic development and food security, they also produce environmental, social, and human health impacts. A World Bank review of the impacts and economic benefits of 50 large dams concluded that these projects showed proven economic and development benefits but had a mixed record in terms of their treatment of displaced people and environmental impacts. A further review by the World Commission on Dams on the performance of large dams showed considerable shortfalls in their technical, financial, and economic performance relative to proposed expectations, particularly irrigation dams, which often have not met physical targets, failed to recover cost, and have been less profitable than expected.

In Pakistan, for example, the direct benefits from irrigation made possible by the Tarbela and Mangla dams are estimated at about $260 million annually, with the farmers who own irrigated land clearly benefiting from increased incomes. However, the increased use of irrigation water has led to waterlogging and increased soil salinity in the Punjab area, with a direct link to a decline in crop productivity, and an increase in malaria transmission.

Hydroelectricity is another important benefit from dams. Total production of hydropower reached 2,740 terawatt-hours in 2001 or 19% of global electrical production, and many industrial (such as Norway and Iceland) and developing countries rely on dams for more than 90% of their power production. As with irrigation dams, in many circumstances the effectiveness of large dams for hydroelec-tricity generation has not been sufficient to meet the predicted benefits, and they have caused loss of habitats and species as well as the displacement of millions of people.

Flood control continues to be another major objective for building large dams. In Japan, for example, 50% of the population lives in flood-prone areas, and in the last 10 years floods have affected 80% of municipalities in the country. Japan is one of the top five dam-building countries in the world. Matsubara and Shi-mouke Dams on the Chikugo River in the Kyushu District in southern Japan, for instance, were built for flood control after a flood in 1953 inundated one fifth of the entire catchment, killing 147 people and destroying 74,000 households. These two dams successfully reduced peak flows in the river years later during a 1982 flood, saving lives and property.

However, the effectiveness of large dams to replace the role of natural wetlands for flood mitigation is not well supported by scientific evidence. Wetlands and floodplains act as natural sponges; they expand by absorbing excess water in time of heavy rain and they contract as they release water slowly throughout the dry season to maintain streamflow. The large-scale conversion of floodplains and wetlands (some of it through dams) has resulted in declines in the natural mechanism for flood regulation. And while a handful of dams are being decommissioned in some countries (268 out of 80,000 in the United States, for example), an estimated 1,500 dams are under construction worldwide and many more are planned, particularly in the developing world. River basins with the largest number of dams over 60 meters high planned or under construction include the Yangtze Basin in China with 46 large dams, the La Plata Basin in South America with 27, and the Tigris and Euphrates River Basin in the Middle East with 26.The debate on cost, benefits, and performance of large dams continues, but given recent reviews, the traditional reliance on constructing such large operations for water supply is being called into question on environmental, political, and socioeconomic grounds.

Interbasin transfers

Interbasin water transfers represent yet another form of securing water supplies that can greatly alleviate water scarcity. They include any canals, ditches, tunnels or pipelines that divert water from one river or groundwater system to another, typically from dammed reservoirs, and often represent massive engineering works involving both ground and surface waters. Changes to natural surface water hydrographs can be enormous and virtually instantaneous. The Great Man-Made River Project in Libya, for example, transports over 2 cubic kilometers of fossil groundwater a year through 3,500 kilometers of desert to huge coastal storage reservoirs that support 135,000 hectares of irrigable cropland, one third of the country's total.

Two of the world's largest interbasin transfers are the 93% loss of flow (27 cubic kilometers per year) from the Eastmain River and a 97% gain of flow (53 cubic kilometers per year) in the La Grande River, both in Canada. In total, the flow being diverted without return to its stream of origin in Canada alone totaled 140 cubic kilometers a year in the 1980s, more than the mean annual discharge of the Nile River and twice the mean annual flow of Europe's Rhine River. The Farraka Barrage alone diverts over 9% of the Ganges River's historical mean annual flow and over 5% of the flow for the entire Ganges-Brahmaputra basin.

A gigantic diversion project is also under way in China, which proposes to move 40 cubic kilometers per year of water from southern China to the parched parts of northern China, thus connecting the Yangtze River with the Hai, Huai and Yellow Rivers. Three channels, two of which are over 1,000 kilometers long, will be needed for this transfer, which corresponds to 4% of the average flow of the Yangtze River. Developers plan to bring enough water to replenish groundwater aquifers in the north. This withdrawal from the Yangtze, even though it represents only a small fraction of the river's annual flow, will likely still have some effect on downstream ecosystems: sediment loads needed to maintain riparian and coastal wetlands will be reduced, and pollutants will be marginally less diluted, raising their concentration in the Yangtze River's lower reaches. In addition, as water flows north from one basin to another, the introduction of non-native species and the transfer of contaminants could affect native fauna in the receiving basins.

Social effects of interbasin water transfers are complex. Populations in the recipient basin of water transfers gain water for irrigation, industry, and human consumption, all leading to indisputable economic and social benefits. However, those living in the basin of origin (and particularly those downstream of the diversion point) often lose precisely those same benefits, and many times they are displaced to other parts of the country, losing their homes and cultural heritage. While sometimes economic compensation is offered to people displaced by dams, the amounts usually do not cover the potential losses in terms of livelihoods, economic productivity, and cultural and historical heritage.

Resettlement is an issue for water transfers as well as for dams, with many resettled communities suffering from a marginalized status, and cultural and economic conflicts with the population into which they are resettled. The central route of the Yangtze-to-Yellow water transfer in China, for example, will require the resettlement of 320,000 people, each of whom is supposed to receive the equivalent of $5,000 in compensation.

The trade-offs involved in interbasin transfer schemes include both direct societal costs and benefits, as well as those involving ecosystems services and biodiversity. Yet given increasing demands for water in the future, such transfers are likely to remain an important mechanism for alleviating regional water shortages.

Virtual Water in Trade

Virtual water, or VW, refers to the amount of freshwater used during the production process and thus "embodied" in a good or service. While tabulations could be made for any product, VW has been explored mainly from the perspective of crop and livestock production and trade, given the predominance of agriculture in water use globally.

Operationally, VW in agriculture can be defined as the quantity of water used to support evapotranspiration in crops, which are then consumed domestically (as human food or animal feeds) or traded internationally. Additional water to process food products and to care for livestock can also be tabulated, but VW estimates are fundamentally determined by irretrievable water losses through crops. There is a vast mismatch between the weight of agricultural commodities produced and the VW embodied in their production. For example, 1 kilogram of grain requires 1,000–2,000 kilograms (liters) of water, even under the most favorable of climatic conditions, producing 1 kilogram of cheese requires >5,000 kilograms of water, and 1 kilogram of beef requires an average of 16,000 kilograms of water.

Water has been transported in internationally traded products for hundreds of years, but the concept of trading VW has only recently begun to be considered as a mechanism to alleviate regional or global water security by exploiting the comparative advantage of water-rich or water-efficient countries. However, VW does not take into account the nature of food production systems and other factors, such as soil erosion, biodiversity impacts, or pollution. Moreover, for political and social reasons, countries may elect to be self-sufficient and independent in food production. For example, India, which is food self-sufficient in aggregate, serves as a net exporter of food and virtual water despite being water-stressed.

A substantial volume of VW trade in food commodities has nonetheless been taking place. Worldwide, international VW trade in crops has been estimated at between 500 and 900 cubic kilometers per year, depending on tabulations made from the exporting or importing country perspective and the number of commodities considered. An additional 130–150 cubic kilometers per year is traded in livestock and livestock products. For comparison, current rates of water consumption for irrigation total 1,200 cubic kilometers per year, and taking into account the use of precipitation in rain-fed agriculture as well, the total water use by crops has been estimated to range from 3,200 to 7,500 cubic kilometers per year, depending on whether allied agroecosystem evapotranspiration is included. The most important exporters of crop-related VW are the OECD and Latin America, though individual sub-regions, such as Western Europe, are net importers of VW. Asia (Central and South) is the largest importer of VW.

Of the top 10 virtual water exporters, 7 countries are in water-rich regions, while of the 10 largest importers of VW, 7 are highly water-short, indicating a general redistribution of VW from relatively wet to dry regions. However, the notable absence of clear-cut relationships linking the degree of domestic water scarcity to dependence on external VW supplies suggests that an optimal redistribution of water through crop production and trade is yet to emerge. The consequences of food self-sufficiency thus entrench present-day patterns of water scarcity, as can decisions to pursue an aggressive export marketing strategy in the face of unsustainable water use. Future increases in water stress over the coming decades and further integration of a global economy are likely to be powerful forces in adopting the notion of VW into food production and trade policies.

Land Use and Land Cover Change

Among the major processes influencing water quantity and quality at the river basin scale are changes in land use intensity and land cover change. Land use changes affect evapotranspiration, infiltration rates, and runoff quantity and timing. Particularly important for human well-being are contrasting reductions in the overall quantity of available runoff with some types of land cover change versus concentrated peaks of runoff associated with flooding under other land cover changes that can often be translocated far downstream through river networks.

For example, expanding impervious areas due to urban expansion greatly increases the volume and rate of stormflow into receiving streams. Such changes also affect the water quality and biodiversity of freshwater ecosystems. Land use changes that compact soils and reduce infiltration are associated with deficiencies in groundwater recharge and dry period base-flow, the long-term global consequences of which are yet to be documented. Reduced infiltration can also lead to longer lifespans of pools with stagnant water, thus providing increased breeding opportunities for mosquitoes and other vectors of human disease.

The impact on local water budgets of changes from forest cover to pasture, agricultural, or urban land cover are well docu-mented in the hydrological and ecological literature. While historically a large portion of the available information was generated for temperate and boreal areas of North America and Europe, information is becoming available for selected sites in Amazonia, South Africa, and Australia, among others. The global impact of 110,000 square kilometers per year net deforestation on runoff, however, has yet to be fully quantified.

Impacts of land use change patterns of weather and climate at different scales are only starting to be understood. Fragmenting a landscape alone can generate changes in local weather patterns. At the continental level, land use changes can reduce recycling rates of water leading to reduced precipitation and distortions in the atmospheric circulation patterns that link otherwise widely separated regions of the globe. There has also been continental-to-global-scale acceleration in the loading of pollutants, including nutrients, onto the land mass associated with industrial agriculture, urbanization, and grazing. These inputs are translated into greatly elevated fluxes to and transport through inland water systems, the effects of which pass in many cases fully to the coastal zone.

Intensive agricultural and urbanized areas have expanded rapidly in the last 50 years. The current extent of cultivated systems provides an indication of the location of freshwater ecosystems that are likely to experience water quality degradation from pesticide and nutrient runoff as well as increased sediment loading. It shows, from a drainage basin perspective, the distribution and pattern of urban areas, as judged by satellite images of nighttime lights for 1994–95. Because more urbanized river basins tend to have greater impervious area as well as higher quantities of sewage and industrial pollution.

Intensively cropped lands are concentrated in five areas: Europe, India, eastern China, Southeast Asia, and the midwestern United States, with smaller concentrations in Argentina, Australia, and Central America. Africa is striking for its lack of intensively cropped land, with the exception of small patches along the Nile, on the Mediterranean coast, and in South Africa. This reflects the minimal use of chemical inputs and the low level of agricultural productivity in most African countries.

The implications of these changes and the incomplete understanding of their consequences affect the manner in which humans interact with the water cycle. Integrated watershed management is the current paradigm for sustainable water use and conservation. It can yield important environmental and social benefits, as shown by a survey of 27 U.S. water suppliers that found the cost of water treatment in watersheds forested 60% or more was only half that of systems with 30% forest cover.

In practice, the integrated management approach is complex and difficult to implement because of limits to the understanding of interactions linking the physical and biotic processes that control water quantity and quality. Integrated management research typically has focused on local and short time scales and been limited to a very small portion of the world's watersheds. Most of the understanding of watershed dynamics and management principles comes from hydrological research on small watersheds and from studies at the local scale. At present, the longest hydrological studies encompass only the last 20–40 years, but the recent application of GIS techniques facilitates reconstruction of past events to place the impact of contemporary land management into a longer-term perspective.

One significant challenge to both scientific understanding and sound management is that multiple processes control water quantity, quality, and flow regime. The pattern and extent of cities, roads, agricultural land, and natural areas within a watershed influences infiltration properties, evapotranspiration rates, and runoff patterns, which in turn affect water quantity and quality. Additional challenges surround the fact that river basins extend across contrasting political, cultural, and economic domains (the Mekong River, for instance, flows through China, Laos, Thailand, Cambodia, and Viet Nam). Thus, there remains substantial uncertainty about the effects of management on different components of the hydrological cycle arising from the unique combinations of climatic, social, and ecological characteristics of the world's watersheds.

It is widely recognized that while much more information is needed to evaluate the impact of land use and cover change on freshwater provisioning services, integrated watershed manage-ment—despite its present degree of uncertainty—is both possible and would contribute significantly to improved management of water resources.

Climate Change and Variability

A major and natural characteristic of the land-based water cycle, and hence of water supply, is its variability over space and time. The large-scale patterns of atmospheric circulation dictate the world's climate zones and regional water availability. One particular concern arises from climate change, which in the past has shaped major shifts in the water cycle, such as changes in the Sahara from a much wetter region with abundant vegetation about 10,000 years ago to the desert of today. A changing climate can modify all elements of the water cycle, including precipitation, evapotranspiration, soil moisture, groundwater recharge, and runoff. It can also change both the timing and intensity of precipitation, snowmelt, and runoff.

Two issues are critical for water supply: changes in the average runoff supply and changes in the frequency and severity of extreme events, including both flooding and drought. Both of these changes have been difficult to articulate due to complexities in the processes at work as well as a non-uniform and, in many parts of the world, deteriorating monitoring network.

Shiklomanov and Rodda present a study of continental-scale variations in water supply as represented in the observational record spanning 1921–88. They used data from a total of approximately 2,500 stations, maximizing, to the degree possible, length of record, suitably large river basins, and hydrographs reflecting near-natural conditions. The stations represented <10% of all available records and reflect great disparities in maximum length of record (from 5 to 178 years). (Statistically, the optimal record length for trend analysis is on the order of 30 years, but detectability of a trend also depends on the relative lengths of the "base" (pre-change) and changed periods of record.)

Year-to-year variations over five continents were 10% or less but rose to as high as 35% when examining 27 climate-based subdivisions. Relatively dry periods occurred in the 1940s, 1960s, and late 1970s, with global runoff declining by up to 3,000 cubic kilometers a year. This is in contrast to relatively wet conditions in the 1920s, late-1940s to early 1950s, and mid-1970s. Though there are limitations to making such global statements, the overall conclusion with respect to renewable supplies of runoff is that despite some recent continental-scale trends (an increase in South America and decrease in Africa), there was no substantial global trend in renewable supplies of runoff over the 67 years tested. Labat et al. did, however, compute an increasing global trend in runoff. This was correlated to increasing global surface air temperature, amounting to 4% per degree Celsius over the last century, though with regional increases (Asia, North and South America) and decreases (Africa) or stability (Europe) over the last few decades.

Care must be exercised in interpreting such long-term trends, which are anticipated to be associated with climate change. Maps of trends presented by the IPCC show large-scale and spatially coherent increases as well as decreases in precipitation over multi-decadal periods that start in 1910, although these patterns shift depending on the time frame observed. A similar time dependency is evident in interpreting changes in rain-to-snow ratios across Canada, with a time frame of 1948–96 indicating completely opposite results than with a time series starting at 1960 and ending in 1990.

The clearest signatures require long time periods and sufficient spatial integration units (that is, large drainage basins). Peterson et al., for example, found it impossible to detect a coherent trend in runoff without first aggregating the flow records from six large Eurasian rivers and over 65 years. Insofar as northern Eurasia is among the regions historically to show the clearest trends in climate warming and the general absence of other confounding effects such as land cover change and water engineering, these results point to the difficulty in assessing recent runoff trends.

Nonetheless, there is evidence that climate change may already be causing long-term shifts in seasonal weather patterns and the runoff production that defines renewable freshwater supply. Shifts toward less severe winters and earlier thaw periods in cold temperate climates that depend on snowfall and snowmelt result in important changes in water availability. Multi-decade hydrological anomalies are apparent for Africa, with decreases on the order of 20% between 1951 and 1990 for both humid and arid zone basins that discharge into the Atlantic.

In the Sahel, persistent rainfall deficits could entrench desertification through a critical loss of water recycling between land and atmosphere, exacerbated by reduced soil infiltration when so-called hydrophobic soils are created in arid environments, and by soil compaction over poorly managed lands. Such rainfall deficits also reduce replenishment of the groundwater resource, exacerbated by the decreased permeability of soils that favor storm runoff and flooding, even in the context of lower overall precipitation. In the transition zones between wet and dry regions across Africa, there is a highly uneven and erratic distribution of rainfall and river corridor flow. While this climate already produces chronic water stress, episodic droughts greatly increase the number of people at risk.

An intensification of the water cycle, through more extreme precipitation in the United States and other parts of the world has also been recorded. However, the effect of these increases on the rest of the hydrological cycle is only now being articulated.

In the United States, where sufficient records are available, Lins and Slack and Douglas et al. used stream gauging stations with 50 years of continuous records (from unregulated systems) to conclude that annual minimum and mean flows have increased. This was later confirmed by McCabe and Wol-lock, who found statistically significant increases in annual moisture surplus (moisture that eventually becomes runoff) over the contiguous United States as a whole, but especially in the East. And while Yue et al. found similar increases in minimum and mean daily flows in northern Canada, they found the opposite to be true (significant decreases in minimum, mean, and maximum daily flows) in the southern part of the country.

Groisman et al. reported that warming in the northern half of the coterminous United States was related to a reduction in the extent of springtime snow cover and to the earlier onset of spring-like weather conditions and snow retreat. This has resulted in the increased frequency of cumulonimbus clouds and in a nationwide increase in very heavy precipitation. Warming in the southwest and northeast part of the country has led to greater summer dryness and increased fire danger. An interseasonal shift of precipitation from summer to fall in the Southeast was also noted.

The effect of increased precipitation extremes on floods is still debated because flood response is influenced by many interacting factors, such as basin geology, terrain, and land cover as well as basin size and rainfall patterns. Also, the natural variability of flood flows can mask small changes in precipitation inputs.

Trends are also apparent in soil moisture distributed around the globe. Historical time series from more than 600 sites indicate a modest increase in growing period wetness for the majority of stations examined, contrary to the expectation (by general circulation models) of drier conditions in mid-continental areas due to climate change.

Taken together, these results indicate a high natural degree of variability and difficult-to-interpret shifts in runoff generation associated with historical climate change. The detection of such changes is complicated by the interactions among existing physical climate variations (that is, decadal and ENSO-type oscillations), land cover change, and water engineering, which for many parts of the world dominate the character of renewable water supplies.

Urbanization

During the twentieth century, the world's urban population increased almost fifteenfold, rising from less than 15% to close to half the total population, and by 2015 nearly 55% of the world will live in urban areas. In developing countries alone, the proportion of the population living in urban centers will rise from less than 20% in 1950 to 48% in 2015. In fact, 60% of the fastest-growing cities with more than 750,000 people are located in the developing world, mostly in Asia. While 70% of the world's water use is for agriculture, the remaining withdrawals are for domestic household and other urban uses, including industry, and in many places these water resources are heavily polluted and limited by local shortages and distribution problems.

Urban residents bring with them a set of new challenges for water supply delivery, management, and waste treatment. Because of the rapid rate of increase in cities around the world, water infrastructure is practically unable to keep apace, especially in the megacities with more than 10 million people. Large parts of these megacities lack the basic infrastructure for drinking water and sanitation, and most large cities in the developing world, and many in the industrial world, lack basic waste and storm water treatment plants.

The geographic location of many of these large and growing cities, such as close to coastal areas, and their rapid pace of growth has encouraged the overtapping of water resources that are not necessarily renewable, such as coastal aquifers. In Europe, for instance, nearly 60% of the cities with more than 100,000 people are located in areas where there is groundwater overabstraction. Groundwater overexploitation is also evident in many Asian cities. Bangkok, Manila, Tianjin, Beijing, Chennai (formerly Madras), Shanghai, and Xian all have registered a decline in water table levels of 10–50 meters. These high levels of abstraction in many cases are accompanied by water quality degradation and land subsidence. For instance, the aquifer that supplies much of Mexico City had fallen by 10 meters as of 1992, with a consequent land subsidence of up to 9 meters.

Overabstraction is also an increasing problem with tourism-associated development, particularly in coastal areas. Ground-water overabstraction in such areas can reverse the natural flow of groundwater into the ocean, causing saltwater to intrude into inland aquifers. Because of the high marine salt content, even low concentrations of seawater in an aquifer are enough to make groundwater supplies unfit for human consumption. Of 126 groundwater areas in Europe for which status was reported, 53 showed saltwater intrusion, mostly of aquifers used for public and industrial water supply.

Unfortunately, the poor, mostly migrant workers from rural areas suffer most from reduced quality or quantity of water supply when they resettle to large cities. Poor residents of cities tend to concentrate in the outskirts, where safe drinking water and sanitation are less available, and they often depend on contaminated sources of water or intermediate water vendors who charge exorbitant prices.

In the context of these many problems, an emerging trend toward protecting water supplies for urban areas is noteworthy. A study of more than 100 of the world's largest cities, for example, found that more than 40% rely on runoff-producing areas that are fully or partially protected. This reflects a growing recognition of the value of ecosystem services linked to sound watershed management approaches, as well as of the limits placed on urban water supply from polluted upstream source areas. The geography of downstream populations supported by upstream runoff-producing areas suggests the potential global importance of this management strategy.

Industrial processes, which include withdrawals for manufacturing and thermoelectric cooling, today use about 20% of the total freshwater withdrawals, which has more than doubled between 1960 and 2000. Even though this global use remains small in comparison to water used for agriculture, the current trend in shifting the manufacturing base from industrial to developing countries, due to globalization and international trade, is of concern for future water security. Much of the technology developed for industry is adapted to industrial nations, which are generally considerably more water-rich. When industrial plants are relocated to developing countries, many of which are water-poor or have limited water delivery services, these operations add pressure to the water resource base and increase conflict among water users. In addition, the environmental safeguards for effluent treatment are less well established or enforced in developing nations, adding to the scarcity problems by increasing pollution.

The most polluting industries, in terms of organic water pollutants, are those whose products are based on organic raw materials, such as food and beverage, paper and pulp, and textile plants. Power station electric generation is the largest source of thermal water pollution. Estimates correlating water withdrawals for industrial use with population density by river basin show that many already water-stressed river basins are also centers for industrial production, such as in eastern China, India, and parts of Europe.

Industrial emissions are released not only as thermal and chemical effluents into rivers and streams but also as gases and aerosols into the atmosphere. These can be transported for large distances and may end up deposited in other water bodies far from the emission source. Large areas of the continents show atmospheric deposition as the single most important source of nitrogen loading, with concomitant increases in pollutant transport through inland waterways.

Freshwater Provision: Consequences for Human Well-being

Water is essential for human well-being, but not all parts of the world receive the same amount or timing of available water supplies. Some areas contain abundant water throughout the year, others have seasonal floods and droughts, and still others have hardly any water at all. In river basins with high water demand relative to the available supply, water scarcity is a growing problem, as is water pollution. Water availability is already one of the major challenges facing human society, and the lack of water will be one of the key factors limiting development. The socioeconomic implications of delivering, using, managing, or buying water also have impacts on human well-being.

Benefits and Investment Requirements

Over the long term, water use has generally increased geometrically, in line with population growth, increased food production, and economic development, and during the last 40 years, there has been a doubling in the water used by society—from 1,800 to 3,600 cubic kilometers a year. In an aggregate sense, water is a required input generating value-added in all sectors of the economy, and trends in its use can be assessed from its ability to yield economic productivity. In the United States, for example, water productivity measured as GDP per cubic meter of freshwater use rose dramatically between 1960 and 2000 by about 25% per decade, to $18 per cubic meter, in response to shifts in regulation, technology, and restructuring of the economy.

Water provided for irrigation has a particularly important role, being responsible for 40% of global crop production. And despite major challenges in conveying adequate drinking water and sanitation, more than 5 billion people are routinely provided with clean water and more than 3.5 billion have access to sanitation. Further, with continued investments in water infrastructure, much of the world's population has benefited from allied improvements in public health, flood control, electrification, food security through irrigation, and associated economic development. From this standpoint, well-managed water resources have helped promote economic development, which is tied closely to improvements in many aspects of human well-being.

A good example is provided by a recent analysis of the cost-effectiveness of different options to achieve MDG 7 (on access to safe water and basic sanitation). Of five scenarios tested, two considered Target 10—halving the proportion of people without sustainable access to safe water by 2015 and halving the proportion of people without sustainable access to improved sanitation. It was shown that for each dollar invested in both improved water supply and sanitation, a return of $3–34 can be expected. Among the health benefits of achieving the MDG drinking water target was a global reduction in diarrheal episodes of 10%. The economic benefits of simultaneously meeting the drinking water and sanitation MDG targets on households and the health sector amount to $84 billion per year, representing reduced health care costs, value of days gained from reduced illness, averted deaths, and time savings from proximity to drinking water and sanitation facilities for productive endeavor.

Because of the variability of the water cycle, economic benefits often accrue only after substantial investments in infrastructure and operations that stabilize and improve the reliability of water resources. Capital investments in water infrastructure totaled $400 billion in the United States over the last century. When the annual investment in water storage for irrigation globally during the 1990s of about $15 billion is tabulated, an important source of required capital can be seen, which can constitute a major fraction of agricultural investment for many developing countries.

Worldwide, investments in dams have totaled $2 trillion. World Bank lending for irrigation and drainage averaged about $1.5 billion per year from 1960 to 2000, although this continues to decrease from a peak of $2.5 billion in 1975 to its current rate of $500 million. Global costs for expanding irrigation facilities are estimated at $5 billion annually, but rehabilitation and modernization costs on existing irrigation works are estimated at an additional $10 billion or more per year. Although projected funding for economic development and meeting the MDGs for the entire water sector is estimated to reach $111–180 billion a year, current investments in sanitation and water supply total from $10 billion to $30 billion annually. Securing water resources is thus deeply embedded within development investment and planning but incompletely resolved. The private sector, with global revenues today standing at $300 billion annually, is a major player in providing potential investments, as described later.

Consequences of Water Scarcity

With population growth and the overexploitation and contamination of water resources, the gap between available water supply and water demand is increasing in many parts of the world. In areas where water supply is already limited, water scarcity is likely to be the most serious constraint on development, particularly in drought-prone areas. Here scarcity is mapped to issues relating directly to human well-being.

While decreased or variable water supply has sometimes presented itself as an opportunity to develop efficiency-enhancing responses and cooperation, more often it has spawned numerous development challenges, including increased levels of competition for water among people and between people and ecosystems; the use of non-sustainable supplies or development of costly alternatives; limits to economic growth, including curtailment of activities and required importation of food and other water-intensive commodities; pollution and public health problems; potential political and civil instability ; and international disputes in transboundary river basins. These situations arise in part because society has typically managed ecosystems for one dominant service such as timber or hydropower without fully realizing the trade-offs being made in such management. This approach has led to the documented decline in freshwater ecosystem condition, with accompanying consequences for human well-being. The poor, whose livelihoods often depend most directly on ecosystem services, suffer most when ecosystems are degraded.

One of the problems thus far has been the difficulty of relating ecosystem condition to human well-being, particularly from the socioeconomic perspective. An emphasis on water supply, by developing more dams and reservoirs, coupled with weak enforcement of regulations, thus has limited the effectiveness of water resource management, particularly in developing regions of the world.

As a consequence, policy-makers are now shifting from entirely supply-based solutions to demand management, highlighting the importance of using a combination of measures to ensure adequate supplies of water for different sectors, and slowly moving toward an integrated approach to water resources management that is now linked directly to development initiatives. Measures include improving water use efficiency, pricing policies, preservation of environmental flows, market incentives, privatization of water delivery, and public-private partnerships among others.

Human society has relied for decades on economic and social indicators for planning, but in virtually complete isolation of measures depicting the state and trends of ecosystem services.

The need for integrated indicators or indices at the national or regional scale to help donors and decision-makers establish priorities in water resource management is widely acknowledged. Such metrics can also assist in monitoring progress toward sustainable development goals in a systematic manner. Many such tools have been proposed over the last several decades. For example, the Water Stress Index was developed in the 1970s to link population to water resources, and various other indices have been proposed, such as the Stockholm Environment Institute's Water Resources Vulnerability Index. New water scarcity indices capitalizing on geospatial data sets and high-resolution digital representations of river networks can define the climatic and hydrological sources of water-related stress.

One important indicator that combines physical, environmental, economic, and social information related to water availability and use is the Water Poverty Index. The WPI is similar to the Human Development Index but applicable to a more local scale, where the impacts of water scarcity are fundamentally expressed. It measures water stress at the household and community level and was designed to "aid national decision makers, at community and central government level, as well as donor agencies, to determine priority needs for intervention in the water sector".

The WPI reflects an attempt to quantify inequities in water allocation and the inability of the poor to govern access to water. It has five key components, each based on a series of input variables that are weighed and aggregated into the overall index. When an element cannot be measured, proxy indicators are used in its place. The WPI relies in part on standardized data collected for other purposes, and thus can be used in comparative analysis of water stress across countries. For instance, to provide inputs on water management capacity the index uses Log GDP per capita, under-5 mortality rates, and a UNDP education index, all used previously in constructing the HDI. The five components of the WPI and some of its key variables are:

- *Water resources*: The physical characteristics of water availability and water quality. This component includes total water availability, its variability across time (seasonality), and its quality.

- *Access to water*: This includes not only the distance to water from dwellings but, more important, the time spent in collecting water, conflicts over water use, and access to sanitation.
- *Water use*: This represents withdrawals for domestic, agricultural, and industrial purposes. In many parts of the world, small-scale irrigation and livestock are key components of livelihood strategies and thus are tabulated as inputs.
- *Capacity to manage water*: This component is measured in terms of income, education level, membership in water users associations, and the burden of illness due to contaminated water.
- *Environmental integrity*: If the ecosystems that support water delivery are degraded, then provision of water per se plus the many services derived from freshwater systems will be jeopardized. This component evaluates the integrity of freshwater ecosystems based on the use of natural resources, crop losses reported in the previous five years, and household reports of land erosion. Overall, no variables of the actual condition of aquatic ecosystems are included, suggesting a component of the index that could benefit from revision.

The WPI has been tested internationally in 140 countries as well as at the local scale in South Africa, Tanzania, and Sri Lanka. Finland and Iceland were found to score the best, while Haiti and Ethiopia fared worst. The results of the local pilot analysis look promising, but the WPI would benefit from a better incorporation of ecosystem condition and capacity measures. Nevertheless, the WPI is a vehicle for understanding the complex relationship between water services and human well-being. Moreover, as the authors state, it constitutes a systematic approach that is open and transparent to all, allowing incremental improvements to the index to be made through community consensus.

There are also important gender-related issues associated with water poverty. Women and men usually have different roles in water and sanitation activities, and these differences are pronounced in rural areas across the developing world. Women are most often the users, providers, and managers of water in rural households and the guardians of household hygiene. In many parts of the world, women and girls can spend several hours a day carrying heavy water containers, suffering acute physical problems as a result. The inordinate burden of acquiring water also inhibits women's and girls' opportunities to secure an education and contribute to family income.

The Cost and Pricing of Water Delivery

Water users in most countries are generally charged but a small fraction of the actual cost of water abstraction, delivery, disposal, and treatment, and in some countries implicit and explicit water subsidies can reach up to 93%. Moreover, externalities associated with freshwater use, such as salinization of soils, degradation of ecosystems, and pollution of waterways, have been almost universally ignored, promoting current inefficiencies in use and threats to freshwater ecosystems. In general, those with access to abundant or underpriced water use it in a wasteful manner, while many, usually the poor, still lack sufficient access to water resources.

When water is in short supply or when it is polluted or unsafe to drink, the expense of delivering water services can rise dramatically or force curtailment in use. As scarcity increases, the cost of developing new freshwater resources also reflects the need to secure water from sources sometimes at great distances from the eventual user, often involving complex hydrological engineering. Until recently there were few incentives in most countries to use water efficiently. However, increasing costs of water supply, dwindling supplies, and losses of aquatic habitat and biodiversity are increasingly providing incentives to value water as an economic good. In most countries, governments bear the burden of water delivery to users, but maintaining necessary infrastructure and expanding it to reach un-served users or improve the efficiency of water delivery is the exception rather than the norm. Inadequate funding results in a lack of new connections and unreliable service, with serious consequences for the poor, who usually incur higher costs when forced to obtain water from alternative sources.

Water can be priced in a number of different ways, and the past decade has shown the increasing application of several common methods, including flat fees, fixed fees plus volumetric charges, decreasing block rates, and increasing block rates. Some of these measures discourage waste, while others lead to overuse.

The Price of Water and Recent Trends

There are enormous disparities in the price of freshwater supplied to end-users, reflecting a complex interplay among several factors, including proximity to natural sources of sufficient quantity and quality, level of economic development, investments—both public and private—in water infrastructure, and governance. A survey of urban households across the developing world showed water costs from both public and private sources varying by a factor of 10,000, from $0.00001 per liter (for piped supply in Calcutta) to as high as $0.1 per liter (through private water vendors). Even municipal supplies can constitute a substantial fraction of monthly family expenditure—for example, up to 20% in informal settlements in Namibia. An analysis of urban areas in Asia showed that prices charged by informal water vendors are more than 100 times that from domestic connections. In Benin, Burkina Faso, Kenya, Mauritania, and Uganda, household connection fees to piped water supplies exceeded per capita GDP by factors of up to 5:1, rendering these unaffordable.

Cities also have seen a marked increase in the cost of financing new water supplies. In Amman, Jordan, during the 1980s ground-water sources were used to meet water needs at an incremental cost of $0.41 per cubic meter. As groundwater supplies declined, the city began to rely on surface water pumped from a site 40 kilometers away at an average incremental cost of $1.33 per cubic meter. In another example, the real cost of water supply for irrigation in Pakistan more than doubled between 1980 and 1990.

In Algeria, drought during 2000–02 forced cuts in the provision of water supply from municipal networks (access restricted to several hours every two to four days), despite large investments in water supply networks by the Algerian government since 1962. The situation was further exacerbated by lack of maintenance of the network, with water losses through leaking pipes and underpricing of the resource use. Today, price increases and a major facilities upgrade are under way. Many African cities have exhausted and polluted local groundwater supplies, necessitating expensive transport of freshwater from distant suppliers or the need to invest in desalination, which is among the costliest methods of supplying freshwater.

In addition to direct prices paid, additional costs are incurred by the poor provision of water services. Time spent in traveling to supplies, queuing, and transporting water can be a significant burden on household incomes for the poor. Compared with the late 1960s, households without piped water supply in Kenya, Uganda, and Tanzania today spend triple the time each day securing water, an average of over 90 minutes. Public taps are often in short supply, as in many Asian cities, where several hundred people are served by a single source. Further, the true costs associated with water delivery services are amplified by significant health burdens incurred when supplies are insufficient to meet basic needs. In the case of Lima, Peru, a major portion of household income (27%) is represented by medical costs and lost wages from water-related disease, while in Khulna, Bangladesh, an average of 10 labor days per month are lost due illness from poor water provision.

Cost Recovery

The fourth guiding principle of the Dublin Statement on Development Issues for the 21st Century articulated that "water has an economic value" and "should be recognized as an economic good." At the same time, the statement argued that water should be available to all people at affordable prices. After much discussion and controversy, which continues to this day, the Ministerial Declaration from the 2nd World Water Forum established that "the economic value of water should be recognized and fully reflected in national policies and strategies by 2005" and that "mechanisms should be established by 2015 to facilitate the full cost pricing for water services, while the needs of the poor are guaranteed."

Supporters of full-cost water pricing argue that to improve efficiency, the set price of water needs to reflect the cost of supplying, distributing, and treating it. There is some evidence that this principle works. For instance, price increases for water in Bogor, Indonesia, reduced domestic consumption by 30%. Proponents of full-cost water pricing also point out that most of the poor are not meeting their basic water needs today under current public management, usually because of lack of government capacity and resources. Consequently, poor communities are already paying higher prices through intermediate water vendors than if they were connected to a water delivery system.

But while pricing water to reflect its true cost is relatively simple in theory, the political and social obstacles are formidable. Opponents to the idea of full-cost water pricing claim that access to water is a fundamental human right. Water, like air, should therefore not be treated as an exchangeable, marketable commodity, because if market conditions rule, access to water becomes dependent on the ability to pay and not an inherent entitlement. In the eyes of many, establishing a price for water or privatizing its delivery puts many of the poorest, most marginalized people at risk of not getting enough water to meet basic needs.

The majority of OECD countries have adopted or are adopting, as an operating principle, the full-cost recovery concept in water management, although what should be covered under this "full cost" is still a matter of debate. Infrastructure costs, however, are not usually included. As pricing was restructured and subsidies reduced during the 1990s and in the current decade in industrial countries to capture the full-cost recovery of water, the real price of water was increasing in 18 out of 19 countries surveyed (Australia being the only exception). In two thirds of OECD countries, over 90% of single-family homes are currently metered.

The concept of full-cost water pricing in the developing world has been introduced with the support of local communities in situations where a more reliable service is assured. In Haiti, for example, shantytown residents with no connection to the water utility pay 10 times more for water from water vendors (water trucks) than those who are connected to the private water utility grid in nearby villages. Residents connected to the grid have their water use metered and pay the corresponding fees.

Water Privatization

One of the most controversial trends in today's globalized economy is the increasing privatization of some water management and delivery services. In many countries, due to increasing costs of maintaining and expanding water networks and overstretched government budgets, private companies have been invited to take over some of the management and operations of public water systems. Private-sector investment in theory results in more financing for infrastructure as well as more-efficient operations and cost recovery, and the hope is that the public will benefit from a more stable and reliable water delivery system at a reasonable price.

Opponents to privatizing water services argue that putting private companies in charge of water will drive prices to the point that marginalized groups have no capacity to secure sufficient water even for their most basic of needs. In addition, because the profit motive fails to recognize environmental externalities, they argue that privatization will increase risk to the very ecosystems that help supply freshwater. The debate on public-private partnerships for water management was prominent on the agenda of the World Water Forum gatherings (in particular, at the 2nd Forum in The Hague and the 3rd Forum in Kyoto), as well as at the World Summit on Sustainable Development.

Despite trends toward privatization, at present over 80% of the world's investments in water, sanitation, and hydropower systems are by publicly owned bodies or international donors. Therefore, the responsibility for providing water, over the short to medium term, will remain largely a public enterprise. Among industrial countries, there is much variation in the degree of privatization. In the United States in 2000, private companies provided only 15% of municipal water supply, although in the nineteenth century they provided nearly 95%. France, in contrast, shows more than half of all residents currently served by private companies.

In Latin America, Chile has been successful at delivering water through privatization, and nearly all houses in Santiago have access to clean water and sanitation. Despite exchange-rate fluctuations, a foreign company, Suez, has remained the water provider for Santiago and its region, investing over $1 billion in water infrastructure. Water in Chile has been priced at rates affordable by the middle classes, and stamps are given to poor people to guarantee near-universal access. Conversely, in Argentina, what looked like a positive trend did not withstand economic troubles. In 1993 privatization in Buenos Aires increased the share of residents served with water from 70% to 85%—an increase of 1.6 million people, with a concurrent drop in prices. Exchange-rate fluctuations in many developing countries, such as the currency devaluation in Argentina in 2002, add challenges to successful implementation of privatization schemes. If the currency of a country devalues, the price paid for water will be worth much less, and the foreign firm could pull out of the market, leaving users without reliable service.

South Africa is using a different pricing scheme to improve poor people's access to water and has made good progress in providing water to nearly two thirds of those who lacked access in 1994, when apartheid officially ended. Despite the relatively low cost of water, however, some rural residents opted to consume free—but contaminated—water from other sources. In February 2000, to improve public health for the poor, the government introduced a scheme to provide households with 6,000 free liters of water per month, enough to provide 25 liters per person per day, with charges for additional use.

Privatization can be executed in many ways, depending on the level of transfer from public to private hands. Full transfer of ownership and operations of water resource systems so far has been rare. The majority of cases embody the transfer of certain operational aspects, such as water delivery, but the ownership of the water resources usually remains with the state, thereby forming a public-private partnership.

These partnerships have been demonstrated over the last few years to capture the benefits of privatization without all of the risk. They do not privatize all of the water assets, but they do give private actors control over some elements of the water rights, infrastructure, and distribution systems. Yet public entities typically maintain ownership over some or all of these systems. Public-private partnerships work best when strong regulatory controls exist. A typical arrangement in France, for example, delegates the operation, maintenance, and development of public potable water and sanitation to private companies, though public bodies retain ownership of the system.

Experience has shown that a clear legal framework, where risks are decreased and the cost of capital decreases, would be necessary to enlist private-sector involvement. More detailed analysis of privatization as a response option for the sustainable management of water resources and freshwater ecosystems is presented.

Consequences of Too Much Water: Floods

In addition to water scarcity, the accumulation of too much water in too little time in a specific area can be devastating to populations and national economies. According to the latest *World Disasters Report,* on average 140 million people are affected by floods each year, more than all other natural or technological disasters put together. Between 1990 and 1999, there were over 100,000 people reportedly killed by floods. The majority of these deaths were in Asia, followed by the Americas, Africa, and Europe.

In addition to human lives, floods are a costly natural hazard in monetary terms, with more than $244 billion damage from 1990 to 1999, the most of any single class of natural hazard. Although this arises from potential changes in climate variability and extreme weather, humans also play an important role, settling and expanding into vulnerable areas.

While catastrophic flooding has negatively affected society for thousands of years, naturally occurring floods also provide benefits to humans through maintenance of ecosystem functioning such as sediment and nutrient inputs to renew soil fertility in floodplains, providing floodwaters to fish spawning and breeding sites and helping to define the dynamics to which coastal ecosystems are adapted. Although floods are primarily natural events, human activity influences their frequency and severity. By converting natural landscapes to urban centers, deforesting hillsides, and draining wetlands, humans reduce the capacity of ecosystems and soils to absorb excess water and to evaporate or transpire water back into the atmosphere, creating conditions that promote increased runoff and flooding.

There are, then, potentially costly consequences of upstream anthropogenic activities on hydrological function that place downstream populations at risk, sometimes affecting other nations, as in the case of more than 250 international river basins. Douglas et al. reported on a simulation study suggesting that, in aggregate, a 32% conversion of forests to agriculture across the pan-tropics has led to a mean increase in annual basin yields of approximately 10%, with a concomitant rise in seasonal high flows. More than 800 million people live along floodplains in river basins containing some amount of tropical forest, and if the most threatened of the existing forests are converted to agriculture in the future, approximately 80 million of them could be at risk from the hydrologic impacts associated with these land conversions. Costa et al. present empirical evidence that large-scale savanna clearance in the Tocantins basin in Brazil has been associated with increases of 24% in mean annual and 28% in wet season flows, independent of climate variations.

Nevertheless, there is some agreement that the most catastrophic floods in large basins result from storms so large and persistent that peak flows are unaffected by land cover. Further, the proclivities of particular regions to landslides, soil erosion, and debris flows, as in the Himalayas, constitute the dominant source of risk. Thus the costs and benefits of designing interventions to mitigate floods have their limits, and there may be little opportunity to escape potential vulnerabilities to flooding, given current patterns of human settlement in high-risk areas.

These findings should not suggest abandonment of good land stewardship, which yields fundamental benefits in sustaining ecosystem services. But they do argue for clearly identifying the source areas of hazard and designing response strategies to protect life and property. Even when specific and well-established policy goals for watershed protection are formulated, stakeholder interests and sustainable funding issues add to the challenge of designing effective upstream-downstream management strategies.

Consequences of Poor Water Quality on Human Health

Water is an essential resource for sustaining human health, and there is a basic per capita daily water requirement of 20 to 40 liters of water free from harmful contaminants and pathogens for the purposes of drinking and sanitation, which rises to 50 liters when bathing and kitchen needs are considered. Yet billions of people lack the services to meet this need, as documented earlier. Water-related diseases include four major classes: waterborne, water-washed, water-based, and water-related vector-borne infections. Threats to health also arise from chemical pollution.

Water-Related Diseases

As a whole, water-related diseases are a leading cause of morbidity and mortality in many parts of the developing world, with estimates ranging from 2 million to 12 million deaths per year, although monitoring and reporting remain poor in many countries. UN/WWAP reports 3.2 million deaths each year from water-related infectious disease, or about 6% of all deaths. The lack of access to safe water and to basic sanitary conditions also translates into the annual loss of 1.7 million lives and at least 50 million disability-adjusted life years. (The DALY is a summary measure of population health, calculated as the sum of years lost due to premature mortality and the healthy years lost due to disability for incident cases of the ill health condition. The DALY is not only an effectiveness indicator in the economic evaluation of different intervention options but also a reflection of the impact of ill health on the income-generating capacity of the poor.)

The first three categories of water-related diseases are most clearly associated with lack of access to improved sources of drinking water, and in turn to ecosystem condition. Improved sanitation through the safe disposal of human waste is a major development objective that improves the health of those served directly by separating drinking water from wastewater. In developing countries, however, 90–95% of all sewage and 70% of industrial wastes are dumped untreated into surface waters, placing both downstream populations as well as ecosystem functions at risk. The fourth category of water-related disease is associated with ecological conditions that favor disease vector breeding. These may be natural (such as those supporting malaria transmission by *Anopheles gambiae* mosquito across large parts of Africa south of the Sahara) or anthropogenic, through improperly planned irrigation systems, dams, and urban water systems.

Waterborne diseases are caused by consumption of water contaminated by human or animal waste and containing pathogenic parasites, bacteria, or viruses. They include the diverse group of diarrheal diseases as well as cholera, typhoid, and amoebic dysentery. These diseases occur where there is a lack of access to safe drinking water for basic hygiene, and most could be prevented by treating water before use. The World Health Organization estimates that there are 4 billion cases of diarrhea each year in addition to millions of other cases of illness associated with lack of access to safe water. This translates into 1.7 million deaths per year, mostly among children under the age of five. Morbidity and mortality from microbial contamination are orders of magnitude greater in developing countries than in the industrial world.

Water-washed diseases are caused by poor personal hygiene and skin or eye contact with contaminated water; their incidence is associated with the lack of access to basic sanitation and suffi- cient water for effective hygiene. These include scabies, trachoma, and flea, lice, and tick-borne diseases. Trachoma alone is estimated to cause blindness in 6 million people. In addition, the transmission of intestinal helminths is linked to a lack of sanitation facilities and is estimated to account for a global annual loss of over 2 million DALYs.

Water-based diseases are those caused by aquatic organisms that spend part of their life cycle in the water and another part as parasites of animals. As parasites, they usually take the form of worms, using intermediate animal vectors such as snails to thrive, and then directly infecting humans either by boring through the skin or by being swallowed. They include Guinea worm infection, schistosomiasis (bilharzia), and a few other helminths (certain liver flukes of local importance in Southeast Asia, for instance, such as *Opisthorchis viverrini*) that infect humans through either direct contact with contaminated water or the consumption of uncooked aquatic organisms.

Although these diseases are not usually fatal, they prevent people from living normal lives and impair their ability to work. For instance, 200 million people worldwide are infected with schisto-somiasis, of which 20 million suffer severe consequences, with an estimated global annual burden of 1.7 million DALYs. The prevalence of water-based diseases often increases where dams are constructed, because stagnant water is the preferred habitat for aquatic snails, their most important intermediary hosts. For instance, the Akosombo Dam in Ghana, the Aswan High Dam on the Nile in Egypt, and the Diamma Dam at the mouth of the Senegal river have resulted in huge increases of local schistosomiasis prevalence.

Water-related vector-borne diseases are caused by parasites that require a vector (such as insects) to develop and transmit the disease to humans. For example, *Anopheles* mosquitoes are the vectors for a protozoan parasite (*Plasmodium*) that causes malaria. These diseases are strongly ecosystem-linked, in contrast to the other three categories of water-related diseases, where water quality (and to some extent quantity) is the key determinant. Their distribution reflects the distribution of ecosystems suited to the propagation of the vectors.

Vector species, moreover, are highly diverse, so that detailed ecological requirements differ over wide ranges. Anopheline mosquitoes—vectors of malaria (1.3 million deaths a year and an annual burden of over 46 million DALYs), for example—breed in different types of freshwater ecosystems and brackish water coastal lagoons. *Aedes,* vectors of dengue and yellow fever, originally breeding in leaf axils of bromeliads, are cosmopolitan in human settlement areas, where they breed in small water pools. Urban filariasis vectors) breed in organically polluted water. And the blackfly vectors of onchocerciasis breed in oxygenated waters of rapids.

These vector-borne diseases are not typically associated with lack of access to safe drinking water but rather with water management practices in tropical and sub-tropical regions of the world. Several parasitic diseases endemic of tropical regions, such as Rift Valley Fever and Japanese encephalitis, spread easily with the presence of reservoirs, irrigation ditches and canals, and rice fields. In all, more than 30 diseases have been linked to irrigation and paddy agriculture. Consequently, improved water management, drainage, and storage practices can help reduce the transmission risk, particularly in areas where anthropogenic conditions have led to the introduction of these diseases.

Chemical Pollution

Another set of diseases affecting industrial and developing nations alike arises in response to chemical pollution of water by heavy metals, toxic substances, and long-lived synthetic compounds. While evidence of the long-term impacts of chemical pollution can be detected even in the remote Arctic, the impacts on poor populations in developing countries are difficult to identify, given the lack of reliable and comprehensive records. However, exposure to chemical agents in water has been related to a range of chronic diseases, including cancer, lung damage, and birth defects. Many such diseases develop over several years, making the links between cause and effect difficult to establish. On a global scale, the burden of disease from chemical pollution is much lower than from microbial contamination and parasitic diseases, but in some highly polluted regions these risks can be substantial. Exposure to chemical pollutants can also compromise the immune system, rendering people more susceptible to microbial and viral infections. The cumulative and synergetic effects of long-term exposure to a variety of chemicals, especially at low concentrations, cannot be well quantified at present.

Naturally occurring inorganic pollutants constitute a class of chemical pollution with serious long-term health effects. Arsenic, which occurs naturally in some soils, for example, can become toxic when exposed to the atmosphere, as seen in areas with high water abstraction from underground aquifers. Arsenicosis is the result of arsenic poisoning from drinking arsenic-rich water over long periods of time and is a great concern in many countries, including Argentina, Bangladesh, China, India, Mexico, Thailand, and the United States. WHO estimated in 2001 that in Bangladesh alone, 35–77 million people—close to half the population—were exposed to drinking water from deep wells contaminated with high levels of arsenic (5–50 times the limit of 0.01 milligrams per liter recommended by WHO). Arsenic is a carcinogen linked to skin, lung, and kidney cancer, although these diseases can go undetected for decades. In other parts of the world, high fluoride concentrations in drinking water have resulted in long-term effects that weaken the skeleton.

Chronic effects also arise from anthropogenic pollutants such as discharge from mining operations, pesticide runoff, and industrial sources. Long-term lead poisoning from old water pipes, for example, can cause significant neurological impairment. Mercury contamination can also originate from industrial discharge and runoff from mining activities, accumulating in animal tissue, particularly fish.

Nutrient runoff is another concern from the standpoint of human health, especially in light of pandemic increase in loadings to inland water ecosystems, for example, of nitrogen. Although there is no global assessment of how many water bodies exceed the WHO guidelines on nitrate levels, most countries report that nitrates are one of the most common contaminants found in drinking water. Coastal and inland waters in regions with high levels of eutrophication have been observed to often propagate toxic algal blooms (toxic cyanobacteria) that can cause chronic disease. In China, for instance, the presence of cyanobacterial toxins in drinking water has been associated with elevated levels of liver cancer. Excess nitrate in drinking water has also been linked to methaemoglobin anemia in infants, the "blue baby" syndrome.

Discharge from aquaculture facilities can also be loaded with pollutants, including high levels of nutrients from uneaten fish feed and fish waste, antibiotic drugs, and other chemicals, including disinfectants such as chlorine and formaline, antifoulants such as tributyltin, and inorganic fertilizers such as ammonium phosphate and urea. These chemicals can significantly degrade the surrounding environment, particularly local waterways. The use of antibiotics and other synthetic drugs in aquaculture can also have serious health effects on people and ecosystems more broadly. The antibiotic chloram-phenicol, for example, can cause human aplastic anaemia, a serious blood disorder that is usually fatal. While many countries have banned the use of chloramphenicol in food production, the level of enforcement varies considerably. A further risk from antibiotic use is the spread of antibiotic resistance in both human and fish pathogens. The U.S. Center for Disease Control and Prevention reported that certain antibiotic resistance genes in *Salmonella* might have emerged following antibiotic use in Asian aquaculture.

There is also evidence from studies on wildlife that humans may be at risk from persistent organic pollutants and residual material that has the ability to mimic or block the natural functioning of hormones, interfering with natural physiological processes, including normal sexual development. Certain chemicals such as PCBs, DDT, dioxins, and at least 80 pesticides are regarded as "endocrine disrupters," chemicals that may interfere with normal human physiology, undermining disease resistance and affecting reproductive health.

Finally, pharmaceutical products excreted by livestock or humans comprise a set of "emerging contaminants," whose impacts on human well-being, ecosystems, and species are not yet understood. These contaminants are hard to detect with current technologies, but their impact on wildlife are already observed in some parts of the world. In the United States, the first nationwide survey conducted in 1999 and 2000 found hormones in 37%

of the streams surveyed and caffeine in more than half. Just recently, 42% of the sampled male bass in a relatively pristine stretch of the Potomac River in the United States were found to be producing eggs. The exact cause is still unknown, but it is hypothesized that it could be caused by chicken estrogen left over in poultry manure or perhaps human hormones discharged into the river with processed sewage.

Sanitation and Provision of Clean Water

Providing "improved" clean water supply and sanitation to large parts of the human population remains a challenge. The most recently completed and comprehensive assessment of improved water and sanitation concluded that 1.1 billion people around the world still lack access to improved water supply and more than 2.6 billion lack access to improved sanitation, with strong geographic variations. Asia contains two thirds of all people who lack access to improved drinking water and three quarters of those who lack access to improved sanitation. Africa is next most prominent in terms of numbers still awaiting improvements in supply and sanitation. Other continents show much smaller numbers but may have relatively low rates of service, as in Oceania, with less than 50% served for both supply and sanitation.

There has been progressive improvement in the provision of sanitation since 1990, recently prompted by the ambitious target for sanitation of the MDG environmental sustainability goal—namely, to halve by 2015 the proportion of people lacking such service in 1990. Worldwide, the goal was set to move coverage from 49% to 75%, and progress is nearly on track with the interim target for 2002 of 62% nearly attained. Of the nine regions analyzed, however, only four are on track or nearly, while five are behind schedule. The greatest challenge remains sub-Saharan Africa, which met only 4% of a targeted 17% improvement by 2002. Western Asia and Eurasia are less off their targets but have not moved forward. Overall, improvements in sanitation in rural areas have been significantly less than in urban areas, and there has even been a decline in the provision of sanitation in rural areas of Oceania and the former Soviet Union.

The rapid and disorganized growth in cities and peri-urban areas in developing countries is likely to hinder progress toward improved water delivery and sanitation systems. In 2000 alone, 16 cities around the world became megacities, with more than 10 million inhabitants each, housing 4%

of the world's population. Most of these megacities fall within regions already suffering from water stress. In Africa, Asia, and Latin America, 25–50% of the population live in informal or illegal settlements around urban centers where no public services and no effective regulation of pollution and ecosystem degradation are available. Half of the urban population in Africa, Asia, and Latin America and the Caribbean suffers from one or more diseases associated with inadequate water and sanitation.

Even if government or municipal authorities were inclined to expand water and sanitation services to informal urban settlements, the lack of formal land ownership, plot designation, and infrastructure make this very difficult and unlikely. In many countries, water and sanitation authorities are only allowed to provide services and connect households to the water grid if proof of land-ownership is provided. These problems are in addition to the basic inability of slum dwellers to pay for connection charges and monthly fees without subsidies. With urban populations expected to encompass 80% of the world's population by 2030, the supply of water and sanitation to city dwellers is set to become one of the greatest challenges to development.

Trade-offs in the Use of Freshwater Resources

It has documented a growing dependence of human well-being on freshwater, which in turn has promoted a variety of engineering strategies aimed at delivering reliable freshwater supplies. So effective has been the ability of water management to influence the state of this resource that anthropogenic impacts are now evident across the global water cycle. Much of the human influence is negative due to overuse and poor management, which has resulted in human-induced water scarcity, widespread pollution, and habitat and biodiversity loss. The capacity of ecosystems to sustain freshwater provisioning services thus has been greatly compromised throughout much of the world and may continue to remain so if historic patterns of managed use persist.

Sector-specific decisions often drive the nature of human interactions with water, with often unintended or purposefully ignored externalities on ecosystems. There is no shortage of examples. Flow stabilization optimizing hydroelectricity can severely fragment and degrade aquatic habitats and lead to losses of economically important fisheries. Industrial development with poor effluent management can result in severe pollution, leading to the loss of aquatic ecosystem function and biodiversity. Connecting urban dwellers to water supply and sewerage systems without due attention to water treatment, as has been commonplace, results in the release of toxic compounds and waterborne diseases that affect downstream water users. In arid and semiarid regions, decisions to promote national food self-sufficiency can translate into great risk to downstream populations and costly infrastructure, as rivers that normally carry water and sediments nourishing coastal lands and floodplains are diverted onto croplands or stabilized behind dams.

Trade-offs are thus an unavoidable component of human-freshwater interactions. Trade-offs are also inevitable in meeting Millennium Development Goals and other international commitments. To demonstrate this, a heuristic analysis is presented here to explore how emphasis on a particular objective could influence the capacity to attain others. The analysis uses the contemporary setting as its starting point, which is then tracked with respect to the impact of five specific interventions. These correspond directly to major objectives embodied in the Kyoto Protocol (carbon mitigation), the MDGs (poverty alleviation, hunger reduction, improved water services), and the Conventions on Biological Diversity and Wetlands (pragmatic ecosystem maintenance applied to inland and coastal ecosystems).

A non-intervention case (current trends) is also considered, analyzing the implications of allowing contemporary trends to continue. A time frame of approximately 10–15 years is considered, allowing sufficient time for general patterns to emerge. This time frame also is associated with the first targets of the MDGs.

The interventions and their impacts are specifically viewed through the lens of freshwater services and ecosystem maintenance. Thus, for carbon mitigation the positive impacts of expanding hydropower to reduce carbon emissions are considered, together with the negative impacts of flow fragmentation that compromises the normal functions of inland freshwater and coastal ecosystems. To maximize relevancy to the international development agenda, the findings refer to poor countries alone. The interventions and key results are summarized in Improvement is depicted by movement outward to the larger circle. Declining condition is represented by a move inward, and no appreciable change settles on the middle curve.

These experiments are not predictions but instead are thematic devices to demonstrate broad-scale effects that can be supported by findings in this chapter. Although the details could be argued legitimately one way or another, it is the basic character of the response that is sought. Furthermore, as will become apparent, it is the behavior of the full set of experiments rather than individual cases that becomes most instructive.

This scenario shows direct beneficiary effects on human well-being but also sustained and substantial declines in the condition of aquatic ecosystems. On the positive side, there is some alleviation of hunger through increased food production that relies on expanded irrigation and use of agrochemicals; continued improvement to health by way of drinking water and sanitation access; some progress toward reducing poverty; and an expansion of hydropower, which in some parts of the developing world (such as South America) is already an important source of energy, with some beneficiary effects on carbon mitigation.

At the same time, aquatic ecosystems and their biodiversity will be increasingly degraded in this scenario due to the combined forces of industrial, agricultural, and domestic sources of pollution, hydropower with associated flow fragmentation, and habitat destruction. Lack of environmental regulation and enforcement exacerbates the trend. Reduced and highly regulated water flows in rivers continue to decrease the transport of water and sediment to estuaries and coastal wetlands. Food provisioning services, in terms of natural inland and coastal fisheries, are in decline, and freshwater provisioning will continue to be placed in jeopardy by the dual threats of overuse and pollution.

Major supporting and regulating services also continue their decline due to loss of ecosystem function across both inland aquatic systems and their linked terrestrial ecosystems. Particularly relevant to freshwater are losses in flood control (from poor land management, erosion, loss of wetlands), in self-purification potential of waterways (from chronic and acute land-based sources of pollution), and in protection of human health (from inappropriate waste disposal). The links between ecosystem services and human well-being mean that these losses of natural services could ultimately compromise the attainment of important development goals.

There is a growing recognition that maintaining biodiversity and ecosystem integrity will require compromise and trade-offs. A good example is the critical choice between providing water for crop production or for healthy rivers and wetlands. In areas where irrigation and storage reservoirs are upstream of sensitive ecosystems, both livelihoods and environmental integrity can be at stake. One possible strategy to accommodate potential losses in food production and income is by managing basin-wide improvements in water productivity for agriculture through new crop breeding, innovative technologies, and water reuse strategies, all saving water and reducing the need for irrigation and flow stabilization.

While only qualitative in nature, these findings clearly demonstrate the consequences of optimizing one development goal or conservation objective over others. This assessment indicates that there would be substantial inconsistencies in the major development and sustainability strategies should they not become better integrated. The impacts of these conflicts on freshwater provisioning services and ecosystem functioning are likely to compromise the sought-after progress inherent in these same international commitments. It is very certain that the condition of inland waters and coastal ecosystems has been compromised by the conventional sectoral approach to water management, which, if continued, will constrain progress to enhance human well-being. In contrast, the ecosystem approach, as adopted by CBD, Ramsar, FAO, and others, shows promise for improving the future condition of water provisioning services, specifically by balancing the objectives of economic development, ecosystem needs, and human well-being.

References

ADB (Asian Development Bank), (2001): *Water for All: The Water Policy of the Asian Development Bank,* Manila, Philippines.

Brismar, A., (1997): *Freshwater and Gender: A Policy Assessment,* Stockholm Environment Institute, Stockholm, Sweden.

Fraser, A., M. Meybeck, and E. Ongley, (1995): *Water Quality of World Rivers,* UNEP Environment Library report no. 14, UNEP, Nairobi, Kenya.

Gleick, P.H. (ed.), (1993): *Water in Crisis: A Guide to the World's Freshwater Resources,* Oxford University Press, London, UK.

King, J.M, R.E. Tharme, and M.S. DeVilliers (eds.), (2000): *Environmental Flow assessments for Rivers: Manual for the Building Block Methodology,* Water Research Commission, Pretoria, South Africa.

3

Freshwater Resources and Social Security

The discussion presented in this chapter centres on the relationship between social security and the management of freshwater resources. These are complex and only hazily-understood issues, and the specific form these processes take in different places will vary greatly. The characteristics of the freshwater ecosystems of different parts of the world are closely linked to local environmental conditions, while social security issues are part of the wider social, cultural and institutional context of an area. The link between the two is not obvious but is real nonetheless.

A secure and stable social context provides the setting within which good management practices can emerge and the integrity of freshwater ecosystems can be preserved. At the same time, the benefits that these resources provide can be important in creating social security. Conversely, the absence of conditions in which people feel secure can lead to environmental devastation, whilst declining access to freshwater resources (through their degradation or increasing demands) can create conflicts and undermine to integrity of the mechanisms through which social security is created.

There is a need to give structure to these complex issues through the development of a conceptual framework that helps understand the many factors any analysis of these issues needs to take into account. Such a framework is presented in this section. The starting point is to understand what social security is and why it matters for the management of water resources and the ecosystems through which they are available. What is not here is an attempt at a comprehensive review of social security or the institutions that regulate the use of freshwater resources, for such a review would be far beyond the scope of a background paper such as this. Such issues are extremely specific to individual times and places and any generalisation will be inherently simplistic. Instead, we present here an approach that can be applied to specific places and societies, illustrating this approach (where possible) with some examples that are intended to stimulate thought rather than 'prove' anything.

Social Security

Social security is experienced by individuals but produced within societies. It is a set of rights and entitlements, and the social and institutional structures through which these are reproduced and made available, that reduce vulnerability to risks and threats beyond the control of individuals. It has material and non-material manifestations, including both the extent to which people are able to meet their most basic of needs (things such as water, shelter, good health and food) in a secure manner and the freedoms people enjoy from threats of violence, prejudice, oppression and environmental risks. For both the material and the non-material, the greater the security the less vulnerable people are to such trends and shocks.

This is directly reflected in the way that society is organised to manage resources and provide for needs. At all levels from the individual to the international (and in an age of globalisation the latter is increasingly important), people interact with each other and the environment through established and understood rules and institutions. These structures provide the benefits that working together brings and are a means through which any conflicts can be resolved. Their effect is mostly implicit: we are all largely self-regulating in the ways we behave to each other and the environment, and most people feel more secure when they know the 'rules of the game' and feel that their needs and rights are represented in these regulatory processes. Where either individuals or groups break the rules, where they act in ways that hurts others or the resources, these regulatory processes can (or should) provide sanctions that curb their behaviour and proscribe destructive courses of action.

Of course, this ideal is often not true: we live in an unequal world in which individuals and groups are able to impose their wants and actions on others. This can be by brute force: the force of arms or the overwhelming power of the majority imposing on a weaker minority. It can be through the institutions that are set up to regulate society: the effects of laws that define 'ownership' or prescribe what can or cannot be done and the operation of government and other agencies that are meant to implement the rules. It can reflect economic power, where the operation of markets and unequal distribution of wealth provides control for some at the expense of others. Or it can be through the lack of societal institutions: where the actions of some may harm others (perhaps downstream in a catchment) but there are no effective mechanisms to control their actions (or perhaps to even understand the implications of their actions) even where there is a desire to. This absence of fair and effective social structures is the root of both ecosystems degradation and social insecurity. As such, any analysis of the relationship between social security and freshwater ecosystems management must inevitably focus on issues of *power and empowerment*, and build this discussion on an analysis of the societal processes through which individuals and groups lead their lives and are linked to each other.

This in turn is in large part a function of the *societal processes* through which people are able to minimise risks and control their own lives. This obviously matters to people. It also matters to ecosystems, as resource degradation is often the result of responses to risks or the impact of shocks that undermine people's security. These can take many forms. The most overt and catastrophic can be wars or civil strife that destroys both people and environments. Though terrible, however, these are by no means the only forms that *vulnerability* takes. There are many types of vulnerability and conflict that undermine people's social security and affect ecosystems viability. How can we understand them? The best starting point is to understand people's lives, how they are linked to social institutions and how this affects the management of ecosystems. This can be analysed within a livelihoods framework, an analytical model that gives us a starting point. This approach is then developed to understand the relationship between individuals and households on the one hand and the wider structures of society and institutions within which people and families live and social security is channelled.

Livelihoods

The concept of *livelihoods* is gaining increasing acceptance as the basis of approaches to understanding the factors that influence the lives and well-being of people, and especially the poor, of the developing world. The concept itself has many meanings and is, like many new but hazy conceptual bandwagons, amenable to abuse as people try to present the old wine of conventional approaches in the new bottle of fashionable development rhetoric. There is a need to bring some clarity and conceptual rigour to this potentially formidable concept. The model presented here is a contribution to this process.

There are many different definitions of livelihoods, but Carney presents one based on the work of Robert Chambers and Gordon Conway that is better than most:

> "A livelihood comprises the capabilities, assets (including both material and social resources) and activities required for a means of living. A livelihood is sustainable when it can cope with and recover from stresses and shocks and maintain or enhance its capabilities and assets both now and in the future, while not undermining the natural resource base".

The whole concept of livelihoods is based around the dynamics of the means through which people secure a 'living': that is, the goods and services that allow them to survive and, for some, prosper. The second key concept is that livelihoods are complex. There is an increasing recognition that the livelihoods of people (and especially households) in the developing world are based around a wide range of activities: people are not just farmers, or labourers, or factory workers, or fisherfolk. Most families base their livelihoods around complex strategies that seek to maximise the use of the bundle of resources and assets they possess or have access to: 'rural families increasingly come to resemble miniature highly diversified conglomerates'.

The third dimension of the approach is that livelihoods are influenced by a wide range of external forces, both within and outside the locality in which a household lives, that are beyond the control of the family. This includes the social, economic, political, legal, environmental and institutional dynamics of their local area, the wider region, their country and, in an era of increasing globalisation, the world as a whole. Its effects are felt by all, including people living in the remotest parts of the developing world. These external forces are themselves not static; indeed, it is their dynamics, the processes of change in the wider economic, social and natural environment, that create the conditions in which livelihoods change. These changes can be longer-term trends: for example changing attitudes to gender roles in a society or the gradual decline in fish stocks in a lake. They can also be sudden shocks: the impact of a war, a hurricane or the collapse of market prices for a key crop.

In general, the poorer a household is, and in particular the less assets it possesses, the more vulnerable it is to disruption in its livelihoods base from these shocks and trends (what can be called the vulnerability context). Indeed, in a livelihoods context this is almost tautologous: vulnerability is both a condition of and a determinant of poverty.

The strategies that the poor, and others, adopt in the face of such threats are often called coping strategies. In these, the household will seek to deploy the different assets they possess to best effect within the (often limited) range of choices they possess. This set of choices is again conditioned by the wider context within which they live, and in particular by the extent to which they can control the key decisions that affect their lives. This is (or should be) why participation is widely advocated: it is about giving the most vulnerable greater choices to meet the risks they face.

The counterpart of vulnerability is *resilience*: that is, the extent to which livelihoods are able to withstand shocks and still prosper. This resilience, of course, reflect assets (or wealth), but it is also a function of the *diversity* of the livelihood system as a wider range of livelihoods activities and assets will give greater choices and mean that disruption to one aspect of life does not disrupt the whole system. Resilience is also derived from the *social context* in which people live, as the ability to cope with shocks and change frequently relates to the access that social structures and contacts give to livelihood opportunities. In particular, there are often sophisticated sets of mutual obligations (what is sometimes called *reciprocity*) within social groups that govern the types of assistance that people give each other in difficult times. As a rule, the more extensive and established the social context of people, the more resilient both their livelihood and the social structures are to disruption.

This central spine is the flow from the *entitlements* and *access* they possess to the resource endowments in their locality through their *livelihood assets* to the set of *livelihood activities* that generate an '*income*' (both cash and kind).

This income is in turn allocated to saving or investments, that enhance the value of the assets, to pay for inputs (fertilisers, raw materials, labour) that goes into production, to repaying loans or social payments (taxes etc) or, finally, to consumption – *the outcome* – that is, the total set of goods and services that constitute the material fabric of people's lives. Obviously, the greater the income, the more that is left after other obligations are met (inputs and social payments) for either consumption (meeting the needs of today) or investment (increasing the ability to meet needs tomorrow). Of course, other factors contribute to well-being, including the social context within which one lives, a sense of freedom and security and many other non-material factors. Despite this, there is no doubt that the root of the 'development challenge' that we all face is poverty: that is, the inability of people's livelihoods to provide them with an adequate and reliable income to meet their needs now and in the future.

Social and Institutional Processes

The first factor is the *local community*: the social groupings that the individual household lives within. The social and institutional structures of local communities are often extremely complex and locality-specific, but they reflect differing combinations of the place (the locality or neighbourhood) and the people (the kin, religious, ethnic, occupational grouping or other social and economic characteristics) where an individual household lives.

The second conditioning factor that affects livelihoods is the *external context*, the legal, political, social, economic and institutional environment: those factors, in others words, that link people and places into regional, national and global systems. This includes the nature and operation of *government* (which can have both direct effects, such as through agricultural subsidies or health services and indirect impacts through policy and macro-economic frameworks, political climates, etc), the structure and strength of *civil society* (those non-state institutions and organisations that also regulate social and economic processes), the operation of *markets* and so on.

These local and external social structures in many ways define the characteristics of the different parts of the livelihood model. For example, entitlements and access to an area of forest to gather products such as fuelwood and fodder can reflect both the legal and policy framework (which define who owns the forest and what form of external regulation exists) and local customs and traditions concerning what can be gathered by whom. This in turn defines a part of the 'natural' capital in the 'livelihoods assets' pentagon. Similarly, both external monetary policies and financial institutions and local moneylenders define the availability and cost of credit, which is crucial in both determining how much income goes to repay past loans and what credit is available for investments and inputs into production.

Finally, the *vulnerability context* has already been referred to. This is the trends and variability in those factors that affect livelihood processes. Most are not different to the local and external context described above: rather they reflect the dynamics of those contexts. As such, the vulnerability context describes processes that can materially disrupt different aspects of the livelihood process. This can be specific: for example, climate change will directly affect the resource base, with other consequences compounding through the system from there. It can also vary, depending on form or timing. For example, a sudden collapse in market prices for a dominant commercial crop can affect the assets available by making key assets of land and agricultural implements less valuable. It can affect the livelihood activity through leading to a decision to plant something different. Or it can affect income if the price collapse happens after planting.

A dominant approach to natural resources management in recent years has been participatory mobilisation to create community-based institutions to manage common property resources. Initiatives such as community-based management of mangroves in Sri Lanka, St Lucia and elsewhere typify this approach. Their points of intervention and impact can be 'mapped' on the livelihoods model.

- The most complete of such initiatives are based on a re-definition of the *external policy and legal context*, with new laws the change tenure and/or access rights and new mandates for government forestry or environment departments.
- The approach also seeks to change the *local social context*, both through raising

awareness and understanding and by creating new local institutions (community-based user groups or management committees).

The model shows the structure of local society in a village in central Bangladesh. Individuals live in family-based households, with typically 2 to 3 generations under one roof. These households are grouped (socially and spatially) in clusters of up to 10 families living in a 'Bari', which is based on kin and social group lines (including caste for Hindus). The Bari is a key decision-making unit for many aspects of resource allocation, including choices over common property resource management. Baris are grouped into 'Paras': larger groupings that are again socially and spatially clustered, but their internal social structure often contains hierarchies and clear patron-client relationships over land and labour allocations. Some contain distinctive occupational groups such as potters or weavers. A group of Paras form a village, which may have several blocks of housing separated by fields and water bodies. Villages are typically administrative units (and hence the main point of interaction with the state), but also often contain central points where markets, services, etc are available (that is, interaction with the wider civil society). A 'village' is consequently a complex construct, with different social units according to different aspects of life. Understanding where decisions are made and how these interact with local social relations is a key to effective participation and empowerment.

- Through this medium, community management changes the *entitlements and access* of individual households to the mangroves. The effects on this will vary from household to household, depending on existing dependence on the mangrove, but in general it is assumed that more secure rights leads to better management practices.
- The combination of better access to the resource base and new institutions usually has a positive effect on the *natural and social livelihood assets* of the households if the community management is both egalitarian and effective.
- This in turn means that *livelihood activities* such as fuelwood gathering or fishing will be more productive and/or sustainable.
- The sustainable *income* will consequently be improved, with less concerns about gathered different products from the mangroves.
- All of these factors together mean that the household's *vulnerability* to declining mangroves, shortages of fuel and food and dangers of storm damage and other hazards from mangrove destruction are all decreased.
- These reduced pressures can have a positive impact on the *common property resource base and endowments*, with improvements to mangrove condition widely observable where successful community management develops.

As such, the effects of changes to the legal framework and the organisation of community-based groups can be traced through the model to analyse cause and effect. The details of the effects on individual households (or stakeholder groupings within the community) can similarly be mapped, to assess the consequences of variations in needs and participation. The impacts of negative developments could similarly be mapped: for example, what have been the immediate and longer-term impacts of the 1998 floods in Bangladesh on different stakeholder groups.

Finally, each of the boxes in the model could be 'unpacked' – detailed out further to give an understanding of the internal structure and dynamics of that part of the system.

There has been a lot of attention in many development initiatives on the issue of 'institutional capacity building'. What does this mean? This crucially includes the district, sub-district and local levels, the arena of local government where the poor come into direct contact with external agencies.Secondly, these stronger horizontal links are complemented by more complete and different vertical links, with more effective penetration down to the local community level and the move from top-down flows to two-way interactions. In particular, there should be a process of devolution of decision-making authority to lower levels, where the people affected by decisions can directly influence or participate in them.Thirdly, the relative strength of central government ministries is diminished and that of local government and civil society strengthened. This means that there is better scope for decisions to be made at the right level (the concept of subsidiarity) and in particular for institutions to be more open accessible and accountable to local communities and the wider society. It also means that there is better integration of different interests in decision-making, which in turn will reduce the risk of conflicts over resources management and create the structure for the arbitration of conflicts where they do emerge.

A key characteristic of existing patterns of resource allocation and social relations is the potential for overt or repressed *conflict*. Such conflicts can be direct and material: for example disputes over access to common property wetlands between fishermen or herders (who want to retain its existing characteristics) and farmers (who want to drain them for farmland that then becomes a private resource). They can be structural and reflect inherent power relationships between genders or social groups (reflecting, for example, the gender divisions of labour in rural production and household maintenance). An essential feature of effective institutional processes is the capacity to mitigate such conflicts in ways that are *fair*, *transparent* and accepted as *legitimate* by all parties.This process of integration can take many forms, and in particular does *not* mean a prescriptive plan, administered by some sort of super agency, which aims to control all aspects of the process, or even formal organisational links and unified chains of command. Integration is above all about the harmonisation of existing policies, programmes, laws and institutions. It is about moving from traditional sectoral approaches that address just part of the picture to a unified approach that addresses the whole picture. An integrated approach should seek to build on what is there, on the social processes and institutions through which social security is provided and freshwater ecosystems are managed. The key to this is to create a common understanding and to construct processes for consensus building and representation that take account of the needs and interests of all groups within the society.

Management of Freshwater Resources

The discussion so far has concentrated on the social and institutional dimensions of our concerns. The basics of this are familiar to us all – water falls as rain or snow, hits the ground and either stays where it is (in ponds or snow fields or as soil moisture), flows over the surface into the river network or infiltrates down into the water table and sub-surface aquifers. From this initial position it can stay where it is, in a water stock on the surface (lakes, wetlands, etc) or underground, with the length of time it stays (or the residence time) variable. It can evaporate or be taken up by vegetation, pass through them and enter the atmosphere again (evapo-transpiration). Or it can move, following the irresistible pull of gravity to flow over or underground and eventually (sometimes stopping along the way) enter the sea – if it doesn't evaporate or isn't taken up by a plant on the way.

This dynamic process of flows within a system (a catchment or river basin) is central to understanding the nature of water resources and their links to human needs, freshwater ecosystem dynamics and the inter-relationships between these ecosystems and social security. *Water resources* are more than just water; they are best understood as a range of goods and services that are derived from a variety of different points in the water cycle and satisfy a wide range of wants. Of course, water itself is directly abstracted and consumed by people in their homes or used in agriculture or manufacturing. Much of this water re-enters the hydrological cycle, albeit in a different form (sometimes as water vapour) and place (generally downstream), perhaps with changed chemical characteristics.

Water resources are also other services that water ecosystems and flows provide: hydropower, for transport, for recreation (including the aesthetic value of many freshwater ecosystems) or for the disposal of waste products from homes and industry. Finally, freshwater ecosystems are the source of many valued plants and animals that are gathered for sale or home consumption or as inputs into manufacturing.

There is no simple, direct link between different types of resource and use: in any one place and time different sources of water can be used to meet any one need and, conversely, different needs can be met by any one water source. For example, water for agriculture can come from rainfall, from surface irrigation or from irrigation which taps groundwater reserves. Similarly, a river can provide water for irrigation, drinking water for people and livestock or act as an input into and source of waste disposal for industry and households. These human uses will inevitably occur in parallel with the ecosystem functions of that river in relation to wetlands it is connected to and the life it contains.

These multiple uses of any water source in an area can be incompatible, both in terms of the amount of water they require and in the effects on the resource they have. And to make life even more complex, the resources themselves move, through the water cycle, so that what happens to water in one place has consequences for all stages in the cycle downstream. For example, using a waterway to dispose of industrial effluents can poison it for drinking, agriculture or recreation downstream or the diversion of water for irrigated agriculture can undermine the viability of wetland ecosystems: in these and many other ways using water resources for one purpose affects the other potential uses of those resources.

What is of critical importance is therefore to understand the institutional mechanisms through which different groups gain access to these resources. This is an issue that is really problematic to generalise about, as the nature and functioning of water management institutions is as place specific as the social structures with which they relate. There are a few general trends that we can identify, but what is more important is that the specific relationship between water management institutions and social security in particular places is seen as a key issue for the development of these institutional processes.

This is not the case in many instances. Water management institutions tend to be centralised, technically-oriented agencies based on one or, at best, two specific aspects of water management. Typical examples of this are water supply utilities (whether state owned or private sector), whose prime purpose is to control water to supply it to homes and commercial uses and to manage sewage disposal systems. Similarly, irrigation agencies are mostly concerned with the channelling of water to farms, whilst hydropower companies manage water to maximise power generation. These and similar organisations are usually engineering-oriented and centralised in operation. They seek to maximise supply with, in many cases, little regard for the impact of their actions on equity or social security (and too often with a poor understanding of their impacts on ecosystems integrity).

There have been attempts to set up more comprehensive agencies, with in particular a lot of faith placed in river basin authorities and planning such as the Tennessee Valley Authority, the Zambezi Action Plan and the Mekong River Commission. These initiatives are usually mandated to take a more comprehensive approach to the management of catchments, but have rarely been effective in integrating the interests and needs of all parties (and especially the poor) in the approaches they take. They tend to be, yet again, technically-oriented and frequently get bogged down in conflicts over the technicalities of water allocation between constituent parts. This is especially true where more than one country is involved, but is also the case (as in India, Australia and the United States) where different states within a country are included and these states have a constitutional mandate over water resources management.

What none of these institutions do is provide a forum for the type of representative consensus building identified here as crucial if existing or potential conflicts between different uses of water resources are to be avoided. The characteristics of water resources identified above (and especially their multiple uses and movement within the water cycle) mean that such conflicts are inevitable. This means that some new thinking is needed on how to develop effective institutional processes for the management of freshwater ecosystems. An interesting, if as yet unproven approach being developed by the Centre for Science and Environment in India is the idea of "River Parliaments": the establishment of a forum in which all interests are represented, which defines the rights and responsibilities of different parties over the management of the river.

Although it will be extremely difficult to get such an initiative going, the approach is based upon social rather than technical processes and as such offers the possibility to integrate social security issues and the needs of all interests into the management of water resources. It is these types of approach, and not traditional 'expert' agencies that point the way forward.

Freshwater Resources and Social Security

The global debate on the future of water resources has centred on one concept: *scarcity*. It reflects neither the dynamics of the resource nor the wide range of different values it possesses. Scarcity should not be seen as a physical absence of water itself, but rather as a shortfall in resource endowments, the total sustainable value of the different goods and services that water resources (and the ecosystems within which they exist) provide. Similarly, resource degradation is not simply a matter of poor quality, it is a loss of endowments as a result of human use of the resource.

The dependence that people have on water resources varies greatly, as does their ability to gain secure access to the resources. All of us need some water, for domestic use at a minimum, but many livelihoods (and especially those of the poor) depend on access to water resources. This can be a direct dependence on freshwater ecosystems for groups such as fisherfolk and boatmen. It can be linked to water elsewhere in the water cycle; rain-fed farmers being a classic example. The ability to secure access to these resources both reflects social structures and, for people highly dependent on such access, is a key determinant of their social security.

These problems affect different places and uses differently: they are as varied as water resources themselves. People throughout the world have learnt to live with and cope with this variability, often through *coping strategies* which are risk minimising and which are highly adapted to the specific characteristics of their area. These coping strategies can and do fail, however, and are in many cases increasingly under pressure from commercialisation and changing patterns of resource use. As such, it is the failure of these coping strategies, which in turn reflects weaknesses in the societal and institutional context in which they exist, not the variable rainfall or river flow, which is the root cause of the problems people face.

It outline of a conceptual framework that provides a basis for both identifying the nature of the relationship of freshwater ecosystems management to social security and understanding what possibilities exist for improving things where this relationship undermines the integrity of the resources or the viability of livelihoods. The approach is used to analyse the situation the world faces today, consider what the future may hold and identify what types of strategies can secure enhanced freshwater ecosystems management and improved social security.

Water is essential for life and is used for drinking and cleansing. It is a key input in agriculture and, unless harvested and artificially applied, is the limiting factor in plant growth in many regions of the world. Given an ideal distribution, there is ample freshwater on the planet to support significant growth in the global population. However, as a resource, water often occurs in the wrong place, or at the wrong time and in the wrong quality. This uneven distribution means that there are many who do not get enough water for their health needs, food production, economic activities and security.

Gleik has estimated that each person needs a minimum of fifty litres of water a day for drinking, sanitation, food preparation and bathing needs to remain healthy. Gleik has also estimated that in 1990 well over one billion people did not have access to this minimum need for water. Using the United Nations population projections, Gleik has gone on to estimate that by 2000, 2,114 million people will not have access to the basic need of 50 litters of water a day to be healthy. The largest increase in under served populations will be in urban areas of developing countries. Nonetheless, the actual number of rural people without access to water by the year 2000 will still outstrip the urban.

Table 1. Developing Country Needs for Urban and Rural Water Supply 1990-2000

	Population not served in 1990 (millions)	*Expected population increase 1990-2000 (millions)*	*Total additional population requiring service by 2000 (millions)*
Urban	243	570	813
Rural	989	312	1301
Total	1232	882	2114

Failing Water Management Institutions

Population growth in developing countries is putting additional pressures on water resources through the need for food production. The drive for increased food production, through 'green revolutions', has encouraged the rapid development of irrigation systems in many developing countries. The WRI estimates that the area of land under irrigation worldwide has increased over six times in the last century. Eighty per cent of irrigated land is in developing countries (Heathcote 1983) and large proportions of many developing country's water resources are devoted to irrigation: over four fifths of the annual withdrawal of water resources in Asia, Central America and Africa are used for irrigation. Irrigation, as part of a policy drive towards food self-sufficiency and crops for exports, has become the imperative for many developing countries. But, this sectoral approach has often eroded the social security of the populations of developing countries. The drive to increase cotton production in the five countries in the watershed of the Aral Sea has not only resulted in one of the most profound ecological crisis on the planet, it has severely affected the health of the population who once lived on the shores of what was once the fourth largest inland water body in the world.

Table 2. Sectoral Withdrawals of Water by Region

Region	***Sectoral withdrawals (percent)***		
	Domestic	***Industrial***	***Agricultural***
Africa	7	5	88
Europe	14	55	31
North America	13	47	49
Central America	6	8	86
South America	18	23	59
Asia	6	9	85
Oceania	64	2	34

Irrigation is a thirsty technology. Only half the water supplied to agriculture is available for reuse. This compares with 90 per cent from water supplied to industry and homes. Over 70 per cent of global water use is devoted to agricultural irrigation. Between 1900 and 1950, global water use doubled and it is estimated that it will double again by 2000 mainly because of population growth and agricultural use. At a country-wide scale this means that several countries in the arid and semi-arid regions are already experiencing a water deficit. Sixteen will have a water deficit by the 2000 and twenty by 2005.

Even where extraction of water is only a fraction of the available water resource, the use of water for food production, industrial manufacturing and power generation creates localised scarcities of water. This is particularly true in rapidly urbanising and industrialising areas of developing countries. The focus of planning continues to be on the development of large schemes to harvest and transport water to areas where water is needed for economic and agricultural growth. Often these projects destroy the livelihoods of large populations of invariably poor groups and appear to benefit groups that have the political leverage and economic power to reap the benefits from such schemes. This is the case in western India where the huge Narmada River Project is currently being developed.

The Narmada Rivers Project exemplifies an approach of large-scale technical solutions to the collection and redistribution of water which benefits large producers at the expense of smallholders; although a large proportion of water resources in developing countries is used for irrigation, much of this is used by larger farmers in better areas. The large proportion of water withdrawal for irrigation masks the reality that large populations of small farmers in developing countries do not benefit from irrigation schemes. They rely on rain fed agriculture, and often sophisticated methods of water harvesting and storage. These methods, developed over hundreds and even thousands of years, have sustained populations in arid regions.

These three cases, the Aral Sea, the Narmada Project and tanks in Tamil Nadu, demonstrate how a failure of institutions to maintain the integrity of the natural resource base and entitlements to water resources results in increased insecurity and impoverishment. The neglect of traditional systems and the focus on large scale water engineering erodes the social security of populations, often forcing them to become environmental refugees, and destroys freshwater ecosystems.

Often the 'water lords' withhold water from the peasant farmers when it is most needed to irrigate the crops. These farmers are forced into a spiral of poverty, paying for expensive water and losing crops because water has been withheld. Forced into debt, they often loose their land. In this area, the erosion of long-established systems of water management, caused by the inability of state institutions to meet the responsibilities that they have assumed, is undermining the livelihoods of the poor, meaning that they lose the few assets they have, face greater uncertainties and increasing impoverishment.

Social Security and Sustainable Resource Management

In these and other ways, water is an area of conflict at scales from the global to the household. Water resources span international borders, regional and administrative boundaries and are, of course, claimed by different communities and even interest groups within communities. At each of these scales social and institutional networks have been developed to define the rights and entitlements of different groups to water resources, but increasing pressures, changing needs and the erosion of many traditional forms of authority are leading to increasing conflicts over these resources. When these networks do not exist or break down then conflict can lead to crisis and ultimately war.

The central theme of conflict resolution at all the scales is to increase security and reliability of access over freshwater resources. The identification of the institutions (the national state, regional governments, communities, etc) which contest for water and the mechanisms and structures they access, or fail to access, to negotiate for water provides a framework through which it is possible to evolve a useful model of conflict resolution. These are indicated by the arrows pointing away from the ellipse towards the scale at which they function.

The International Level

Water, it is claimed, will be the source of conflicts and war in the next Century as oil was in this. Nowhere is this seen more clearly than in the countries of the Eastern Mediterranean. The Grand Anatolia Project in Turkey threatens the flow of the Euphrates in Iraq and Syria. Egypt has frequently asserted that it will go to war with the riparian states of the Nile if they should ever threaten its sole supply of water. Israel has already used its military might to ensure that plans by the Arab Summit of 1964 to develop the headwaters of the Jordan and divert them were abandoned by occupying large areas of these headwaters. The US intelligence services estimate that there are at least ten places in the world (mostly in the Middle East) where war could break out over the shortage of freshwater resources.

However, there are also many examples of co-operation between riparian states in the allocation and control of freshwater resources. The United Nations compiled a list of 3,707 such agreements between riparian states. The conflict between India and Bangladesh exemplifies the problem of over extraction leading to reduced availability of water resource in the lower riparian state. In 1975 a barrage was completed at Farakka, 18 kilometres from Bangladesh border on the River Ganges. The barrage was built to divert water through a feeder canal to supplement the dry season flow of the Bhagirathi-Hooghly, the river that serves Calcutta and its port and irrigation needs in West Bengal. India diverts 40000 cusecs of water from the Ganges during the dry season. On occasions this has reduced the flow of the Ganges to a mere 7000 cusecs, causing damage to the riparian environment in their state, and in particular affecting the world's largest mangrove forest, the Sunderabans. A number of attempts have been made to resolve this dispute, culminating in the signing of the treaty between the two countries in 1997 that defined the mechanisms for sharing the Ganges waters.

The quality of flow is also a significant area of contention between upper and lower riparian states. The Colorado River shared by the United States and Mexico has been the subject of an agreement since 1944 when the US guaranteed and minimum flow of 1.5 million acre feet annually to Mexico. During the 1960s the water quality deteriorated because of increasing salinity due to drainage from the Wellton-Mohawk Irrigation Project, causing decreased crop production in Mexico. This led the US building a desalination plant and agreeing to deliver to Mexico water that would be no more salinated than 300 ppm. Similarly, polluted with waste salt for potassium mines in the Alsace Region was making the River Rhine unusable for agricultural activities and threatened the fish population. A 1975 political agreement between Switzerland, France, Germany and the Netherlands has co-ordinated water quality data on the Rhine. This action has resulted in a significant improvement in the quality of the entire riparian ecosystem of the Rhine.

Conflict resolution is generally developed without recourse to International Law. Issues of quantity and quality of water resources have been resolved in some, but not all, cases. Indeed, Wallensteen and Swain suggest that many of the bilateral agreements that have been developed in the past are under threat because of the increasing demands placed upon water resources through growing populations and the need for food security. At present no international tribunal exists to mediate in water related disputes. Nor is there a consensus amongst international legislators. "What is needed" McCaffery argues, "is a joint management mechanism, established by agreement, participated in by all riparian states, and perhaps supported by neutral, outside parties, whether states or international organisations".

The National Levels

The lack of management mechanisms to resolve conflict in the international arena is reflected within countries as well. Devolution of responsibility for water management and planning to states within countries means that often there is a need for the development of appropriate mechanisms for resolution within countries. The on-going allocation of rights to water supplies from the Cauvery River between the States of Karnataka and Tamil Nadu in India exemplifies the need for the development of the necessary mechanisms within countries to mediate over water supply. As long ago as 1807 the damming of the Cauvery River in southern India was a source of dispute. The Princely State of Mysore, which controlled the upper reaches of the river, built dams to facilitate irrigation. This led to a decline in the flow of water to the rice bowl of Tamil Nadu, then part of the Madras Presidency. In 1829 the King of Mysore made an agreement with the Madras Government to develop no new irrigation projects without the permission of the Madras State.

But the dispute over the release of water into Tamil Nadu from Karnataka has rumbled on throughout this century. Failed attempts at bilateral agreements in the 1970s culminated in the Tamil Nadu Government going to the Supreme Court to demand a settlement. The Court appointed the Cauvery Water Disputes Tribunal, who reported in 1990. This tribunal allocated 205 tmc of water to Tamil Nadu every year and restricted the expansion of irrigated land in Karnataka. Nonetheless, the supply of water to the rice bowl of Tamil Nadu remains a contentious issue. The new agreement has failed to define when the water will be released and the Tamil Nadu Government is pressing for the development of a schedule of weekly releases to ensure that water is available during the critical summer months to irrigate the *kuruvi* crop. As M.S. Aiyer, a senior Congress (I) politician has noted, "if no water is available in a specified week in June there is no purpose in releasing it in September". The arguments over the reliability of the release of Cauvery waters rumble on.

The characteristics of many national water management institutions, that are typically concerned with one set of issues and that do not effectively represent the interests of all constituencies in their operation. These problems are made worse where, as in India, jurisdictional responsibilities are split between different authorities (in this case states) that have no established platform for collaboration and co-ordination. The discussion on the nature of institutional development emphasised the importance of integration and collaboration, but this in turn may require some fundamental changes to the structure and mandate of different levels of authority within a nation.

Local Level Urban

The world is becoming an increasingly urban place, with the majority of the population of the Developed World and of Latin America already urban dwellers and the cities of Africa and Asia growing rapidly. In most cases, these communities are not directly involved in the management of freshwater ecosystems, but they do need access to the services that these resources provide. For these people, a major concern is consequently the character and effectiveness of the intervening agencies through which these resources are provided.

A key issue at a local level is the negotiation of water supply needs between communities and service providers and within communities. In the slums of Bombay, for instance, there is a relationship between the ability of slum dwellers to organise around the issue of water provision and to negotiate with service providers and the quality, reliability and cost of water provision in the slums. A recent survey in thirty three slums in the suburbs of Bombay established that in slums which have developed effective local organisations water supplies are adequate, while in slums with no organisation water is controlled by private landlords, supplies are unreliable and the water is expensive.

In the first group of slums, ensuring a regular supply of water supply has involved the building and maintenance of local water networks. This building work has involved considerable co-operation between households in the slum and the development of organisational structures to develop and maintain the water system. These organisations must continuously negotiate with the municipal service provider, the Brihanmumbai Municipal Corporation (BMC), to maintain the slum's water connection. The continued supply of water to the slum by the BMC is ensured through regular payments collected by these organisations within the slum to cover water bills and through a recognition by the service providers they enjoy the popular support of the residents of the slum.

Access to adequate and reliable water is reflected in the health of the slum dwellers. In the adequately served slums the slum dwellers report a lower perceived incidence of diarrhoea than in the under serviced slums. It is also reflected in the nutritional status of the slum dwellers. In the adequately served slums none of the children between the ages of one and five years were malnourished, while in the under served slums over 75 per cent of the children were either moderately or severely malnourished. Although many other factors confound a direct correlation between water supply and the nutritional and disease status of the children, in this survey it is apparent that the development of appropriate organisation with which to develop and negotiate with service providers for water is an important aspect of maintaining livelihoods in the slums of Bombay.

Local Level Rural

Social organisation to access and manage water resources is also an important feature of rural life in many parts of the world. Machakos District, situated in south-east Kenya, is considered by many to be an example of what rural development can achieve. With funds provided by the Swedish International Development Agency, the Kenyan Government started projects with traditional local organisations in 1979. These local organisations, called mwethya groups, are mainly women. They practice a terracing method known as fanya-juu. This involves ditch digging along the contours of hills. The soil from the ditch is thrown up the hill to form a bench terrace. The value of this method of terracing is that it maximises erosion control while ensuring rainwater control. Since 1979, over 1,000 kilometres of terrace have been developed by the mwthya groups; 70 per cent of the cropland in the area is now terraced. Postel estimates that corn yields on these terraces have increased by over fifty per cent. Critchley suggests that "the existence of well developed self-help groups is one of the main reasons for the success of conservation activities in Machakos".

Such social structures are important in the normal rhythm of life, but can be of critical importance in times of real jeopardy, when extreme environmental events (or human artefacts like markets or wars) threaten their very existence. The role of social organisation in preserving the collective good and conserving fragile resources is particularly important in areas, such as the Sahel region in Africa, where high levels of environmental hazard or variability are the norm.

These traditional management systems can, however, come under threat as wider economic and political changes penetrate rural areas. These types of trends are found in many parts of the world, and illustrate the importance of understanding the dynamics of change in all aspects of social relations to water resources. It is these dynamics that have the potential to undermine both social security and freshwater ecosystems, but that are also the context within which sustainable management opportunities must be understood.

The influence of external institutions need not be negative, however, for they often possess great potential to be the catalyst for the emergence of more robust social relations and more sustainable resource management approaches that both meet people's needs and conserve the resource base even in the face of, as in the example, the most extreme forms of external threat. This is the challenge, to identify and build on these examples of good, these models of best practice to ensure that the approach to freshwater resources management is one that reduces the vulnerability of both social institutions and freshwater ecosystems to negative impacts from external forces.

Role of Ecosystems in Social Security Provision

At all three scales, the national, regional and community the common theme is that fair and sustainable access to water is defined by the ability to develop appropriate institutions to allocate these resources and to meet stakeholders' perceived needs. Often this process will have to contend with competing interests and agendas. Such an investigation raises questions about reliability and security of access to water resources. As will be noted from the case studies presented above these issues need to address access by countries, regions, communities and households to water to produce adequate food, continue their economic activities and remain healthy. The discussion must also address the sustainability of the uses that water is put too. Human activity can degrade ecosystems that are reliant on water, through changing the quality and quantity of water available to the ecosystem. Over exploitation, either directly or indirectly, imperils the viability of ecosystems which are vital for the Earth's health and threatens the social security of this and future generations.

The World Resource Institute suggest that the real value of ecosystem services is about twice that of global gross national product, about US $33 trillion. The authors point out that human life is impossible without the ecosystem services outlined. Nonetheless, the incorporation of ecosystem services into the discussion making and planning is still rare. Only when degradation impinges directly on human activities is the importance of its value truly appreciated. Often this is valued in the costs incurred in the rehabilitation of ecosystems or in providing alternatives, not the inherent values of the ecosystem itself.

Table 3. Ecosystem Services: Free but Valuable

Ecosystem service	*Value $US Trillion*
Soil formation	17.1
Recreation	3.0
Nutrient cycling	2.3
Water regulation and supply	2.3
Climate regulation (temperature and precipitation)	1.8
Habitat	1.4
Flood and storm protection	1.1
Food and raw materials production	0.8
Genetic resources	0.8
Atmospheric gas balance	0.7
Pollination	0.4
All other services	1.6
Total values of ecosystem services	33.3

As was noted in the discussion of the polluting effects of the Wellton-Mohawk Irrigation Project on the Colorado water uses can have a profound effect on livelihoods and ecosystems. Nowhere is this seen more clearly than in the case of the Aral Sea presented. In both cases the rehabilitation of these freshwater ecosystems is very expensive. On the Colorado River this has involved the building of a desalination plant to ensure that the River's water quality is adequate for other human uses downstream in Mexico. Similarly, the rehabilitation projects initiated in Alsace on the Rhine to reduce salt inputs from potassium mines has cost an estimated 100 million French Francs. A huge rehabilitation scheme is now underway in the five riparian states of the Aral Sea and its water sources funded by the World Bank, UNDP, UNEP and the Global Environment Facility. US$ 40 million has been spent in the preparatory phase alone and a further US$ 260 million is already committed for restoration work in the first phase of the project.

It may be inferred, therefore, that any human activity to use freshwater resources will inevitably degrade the quality of freshwater ecosystems. Certainly from the examples presented above this would appear to be the case. Economic imperatives combined with demographic trends and the need for increased food production and security appear to lead to freshwater ecosystem degradation (the more so given that these ecosystems are rarely valued in economic terms). But this need not be the case. The Mahakali Treaty, was explicitly developed to address issues of river ecosystem maintenance in Uttar Pradesh as well as addressing the economic needs of the two states linked in one river basin.

The challenge is to develop strategies that account for both the maintenance of freshwater ecosystems and address needs for their use. This is a balancing act in which, as has been shown in several of the examples above, the degradation of ecosystems leads to increased vulnerability. Nonetheless, freshwater ecosystems must be exploited. In a world where, as was pointed out at the start of this discussion, over two million people do not have access to sufficient water to be healthy at the turn of the millennium strategies of ecosystem conservation by excluding people from them to maintain their pristine condition does not offer a feasible alternative. This too will increase the vulnerability of populations and is unlikely to be effective in maintaining the ecosystems as encroachment inevitably occurs.

It is apparent that innovative methods will have to be sought to develop a balance between the freshwater resource use inherent in a livelihoods approach and ecosystem maintenance. The experience of New York City to the management of its water resources points to the challenges that face water resource planners. Recent investment in the purchase and management of land in its water catchment has saved New York City the expense of installing expensive treatment plants to purify water for the city. Aside from saving an estimated US$ 6.5 billion, the management of these upland resources has created new wildlife habitats, other ecological resources and provided areas for recreation. This experience points to the need for the sustainable management of resources while still addressing human needs for freshwater resources.

The balance between the two imperatives of freshwater resource management is the challenge that faces us as we enter the 21st Century. As has been dealt with in some detail, there are a large number of institutions at scales ranging from the international to the community which negotiate for freshwater resource use to maintain livelihoods and increase social security. Within these forums there is also a need to address sustainable management practices of freshwater resources. It is to these issues that the discussion now turns in an investigation of the global scenarios for freshwater management in the 21st Century.

Possible Future Scenarios and Social Security

Scenario generation is an important tool for long term planning and policy formulation. Scenario generation involves creating a narrative, using socio-economic and environmental data, to indicate a possible future. The creation of different water scenarios combines many interacting elements including: growth and demographic patterns, economic scale and structure, technology and efficiency, policies and institutions, lifestyles and values. These water scenarios help policy makers and managers understand how the world might change, recognise when it is changing, and if it does change, know what to do.

The data sets behind these scenarios are often incomplete and do not contain sufficient details to accurately predict future outcomes. However, the objective is not to accurately predict (past water predictions have shown the difficulty in doing this, for a discussion of this issue but to create a vision which is aimed at directing future water policy. Therefore, this is a visionary process, one aimed at looking into the future and asking 'what implications does this development trajectory have for global, regional and local water use?' If this is not satisfactory or sustainable, a different vision can be created and in doing this, planners and policy makers can make decisions on how to alter conditions to arrive at the chosen 'vision'. All these scenarios or visions have their basis in current water withdrawal and consumption rates and use today's data as their starting point.

The current socio-economic and environmental situation is dominated by factors of influence known as *driving forces* and it is these forces that will determine the direction of our water future. Driving forces include prevailing economic approaches, human values and beliefs, scientific paradigms and political systems. At this point, the scenario is developed using the current driving forces. The scenario narrative is told through the evolution and development of these driving forces.

Within the scenario there are also *attractive* and *repulsive* forces. Attractive forces are those that suggest possible futures are in line with sustainability principles, such as consumption limits and resource use. Repulsive forces suggest non-sustainable futures and thus are not followed. Finally, *sideswipes*, are those unknown and unpredicted influences that may arise during the development path, e.g. a world war or global epidemic. These driving forces may take us three ways: to the conventional water world (CWW), to a world where there is a water crisis (WC) or to a sustainable water world (SWW).

Projecting future water use has been a preoccupation of scientists for many years. Depending on the overriding scientific and social climate at the time of making these projections (for example, population explosion, acid rain, depletion of fossil fuels or climate change) there have been different levels of assessment. It shows the range and variation in these assessments.

Table 4. Global Water Projections

Study	*World Withdrawal (km^3)*	*Year*
Nikitopoulos	6730	2000
L'vovich	7000	2000
Falkenmark and Lindh	8380	2000
Falkenmark and Lindh	3986, 4961	2000, 2015
De Mare	6080	2000
WRI	4195-4350	2000
Shiklomanov	5190	2000
Shiklomanov	3940	2000
SEI	5000	2025

Presently, it is widely accepted that the assessments made by SEI and by Shiklomanov are the most accurate. However, even the latter has built upon his mistaken predictions to come up with the most recent shown. The assessments made by Shiklomanov in 1987 for the period 1990 to 2000 greatly overestimated global water withdrawal and consumption because he overestimated the growth in irrigated lands and the growth in withdrawal for industry. The growth rate appeared to be not so high in the majority of countries as it had been assumed.

There is a general pattern that assessments are falling, peaking at 8380 km^3 in 1974 and falling to 3940 km^3 in 1997. This reduction may be due, in part, to the realisation by scientists of the capacity of humans to respond to water shortages (or threats of water shortages) through the development of new technologies or changes in consumption patterns (e.g. from water recycling to improvements in irrigation technologies). Also, as threats such as acid rain are dealt with by the international community, the projections will alter accordingly.

Therefore assessing future water use is a dynamic process and will change depending on the overriding social concerns and the level of available technology. An additional problem which relates specifically to the CWW is that this development trajectory will, by implication, alter its own course. The CWW is likely to be knocked off course by the stresses it will impose on environmental resource and institutional systems, and by future surprises that are incompatible with the assumption of evolutionary continuity in socio-economic systems.

However, the responses to these stresses are also not known, so it has to be assumed that the former will guide the CWW back on its assumed trajectory. As long as the scenario process is not considered to be scientifically rigorous, it can help developed and developing countries make plans for the future.

There are differences in how the following three scenario's are presented. The CWW is presented with statistics, which are based on current water withdrawal and consumption rates. The CWW also uses current socio-economic and political structures as a basis with which to predict future arrangements. The other two scenarios can be classified as 'visions' as they are presented as narratives and are based largely on 'what ifs'.

Conventional Water World

The CWW builds up a picture of water use and availability from the assumption of continuing current trends. There is an assumption that demographic, socio-economic and technological patterns gradually evolve without significant surprises, radical technological innovations or fundamental policy changes. The CWW assumes the dominance and evolution of the current development paradigm that encapsulates; globalisation, free trade, private investment, unregulated markets, competition, mid range population growth, urbanisation, industrialisation of the periphery, and governance based on the nation state model. The guiding principles of the CWW are evolution, convergence and integration.

In the CWW, global economic patterns persist and there is a continuing concentration of wealth in the industrialised countries.

Table 5. GDP Projections (billion US$ 1990)

Region	*1990*	*2025*	*2050*
N.America	6040	14884	21063
W.Europe	7171	15917	23660
OECD Pacific	3524	8100	11748
Former USSR	854	1898	2756
E.Europe	210	467	679
Africa	401	1657	4245
S.America	994	3018	6038
Middle East	541	2237	5071
China	451	2698	6391
S and E Asia	1043	4943	12631
World	21230	55820	94282
Industrial	16735	38901	56471
Transitional	1065	2366	3435
Developing	3430	14553	34376

The industrialised nations have considerably smaller populations that the developing economies and this pattern continues with significantly higher growth rates in the latter.

Table 6. Population Projections (millions)

Region	*1990*	*2025*	*2050*
N.America	277	330	322
W.Europe	456	489	477
OECD Pacific	145	161	157
Former USSR	289	332	349
E.Europe	100	115	121
Africa	640	1519	2204
S.America	445	699	812
Middle East	151	384	557
China	1223	1733	1867
S and E Asia	1564	2634	3214
World	5290	8395	10080
Industrial	878	980	956
Transitional	389	447	470
Developing	4023	6968	8654

This means that globally, much of the economic wealth is concentrated in a relatively small proportion of the global population, in the industrialised countries. It is in the countries of Europe and North America where water withdrawals are highest. It is also in the industrialised countries where water will have the highest associated costs.

Table 7. GDP and Water Intensity by Region in 2025

Region	*GDP (10^9 1990 US$)*	*Water Intensity (litres/$)*
N.America	14892	41.5
W.Europe	15916	19.8
OECD Pacific	8099	17.2
Former USSR	1896	258.7
E.Europe	468	177.4
Africa	1657	153.8
S.America	3016	122.5
Middle East	2240	142.5
China	2698	290.7
S and E Asia	4944	330.7
World	55826	89.7

The ability of industrialised countries to recover costs will have high implications for water sustainability. The associated politico-legal structures in these regions will make this exercise easier. In the developing countries however, water is considerably cheaper although access to water and water of quality is problematic. A particularly important trend in the CWW is the rapidly urbanising global population, especially in developing countries.

Table 8. Urban Population, 1950 – 2020 (%)

Geographical Region	*1950*	*1960*	*1970*	*1980*	*1990*	*2000*	*2010*	*2020*
World	29	34	37	40	45	51	57	62
Developed Countries	54	61	67	70	73	75	78	81
Less Developed Countries	17	22	25	29	37	45	52	58
Africa	15	18	23	28	34	41	47	54
S.America	42	49	57	65	72	76	80	83
N.America	64	70	74	74	75	77	80	83
Asia	16	22	23	26	34	43	50	56
Europe	57	61	67	70	73	77	80	83
Oceania	61	66	71	71	71	71	73	76

In industrialised nations, there will not be a significant increase in the numbers of people living in towns and cities and consequently, there will be minimal strain put on urban systems. Access to domestic water in these areas will be almost guaranteed although, as shows, costs will be high. This ensures a vital ingredient for social reproduction in these areas. In developing countries however, the growing urban population above the planning capacity of these areas will have serious implications for social security. Much of the urban growth will be in unplanned settlements with little or no access to safe water supplies or adequate sanitation. This will result in the significant increase in the number of vulnerable people in urban situations in developing countries, with a risk of an increased incidence of water borne epidemics.

Water use, by sector, will be dominated by agriculture, particularly irrigated agriculture, although rain-fed agriculture will experience something of a renaissance. This will be due to a general saturation of productivity levels with irrigated agriculture (particularly in Asia) and on increased reliance on rain-fed agriculture as large sectors of the population in developing countries cannot afford to participate in irrigation schemes.

Industrial water use will also see an increase, almost doubling by 2025. Much of this growth will be in the developing countries, as the water intensive technologies as a means of production which are owned by transnational corporations are "exported" Industrial water use in developing countries will begin to slow down as the 'dematerialization' of production begins. This pattern will increase the low-grade employment opportunities in developing countries and it will mean that more available water resources will be diverted to water intensive industries. This often threatens access to water during spells of drought.

The institutions governing water use are increasingly located in the private sector. This is particularly true in the developed economies where water extraction, supply and maintenance are all in the hands of the private sector, with government providing a regulatory role. This limits the influence of civil society on water use and supply and the relationship is reduced to supply and demand economic rationale. In developing countries the private sector plays an increasing role in the provision of water infrastructure because national governments are increasingly experiencing public sector spending deficits. In cases where governments retain control, the emphasis is on low cost, decentralised, local institutional water management. The process both empowers local communities and reduces their access to water resources as private, economically motivated enterprises dominate the control of water.

Economic processes within developing countries become to echo global economic patterns, with an increasing polarisation of wealth and access to resources within country by a growing elite. This is not echoed in national economic indicators but is seen in poverty indicators. These patterns seem to refute the arguments of those scientists involved in the water predictions who used population growth as the critical variable in their calculations. The increasing concentration of wealth within countries means that a smaller number of people are increasingly using more water and large, growing populations are surviving on less water per capita than previously. This pattern can particularly be seen in urban situations where low density, well kept, water intensive households, back onto high density, unplanned settlements with no access to safe water. This, accompanied by the globalising, neo-liberal economic climate means that the private sector are dominating water extractions which they are using to export both food and industrial products.

This general process results in a growing number of vulnerable people. It is therefore clear that population alone should not have been used as the major variable, rather it is the economically empowered population that have had most impact on water withdrawal and consumption. In this respect developed countries will have much higher levels of social security in relation to freshwater than developing countries. In a similar fashion, the wealthy sectors of developing countries will also have more social security than the poorer sectors (and non-urban) – this also relates to water quality. It is therefore necessary to highlight that regardless of the availability of physical assets, it is the *access* to these assets which is important and much of that access depends upon income levels; thus the polarisation of wealth debate indicates a move towards the smaller more privileged sectors of society, on a global and regional level, having more social security than the larger, underprivileged sectors.

Water Crisis

The water crisis scenario originates from the CWW vision which continues up until 2015 without deviation from its original path. However in this scenario global economic growth does not increase to the extent where convergence of income levels begins. There is more unequal competition for global resources results in increasing polarity of wealth and the 'haves' lives in a world that resembles a rich ghetto in an increasingly barbarous planet. The slower growth of the WC scenario results in a number of nations, particularly those that are already economically disadvantaged, experiencing a fall in standards of living. This increases the number of people living in relative and absolute poverty. As a result widespread environmental degradation occurs, as people are pushed onto even more marginal resources, vulnerability increases to epidemic proportions and migration to the more wealthy areas is often the only option.

Water abstractions slightly increase from the level of the CWW but there are no concurrent developments in efficiency or in a change of consumer patterns. The investment in water infrastructure increases slowly, in line with the general rate of economic growth. Economic expansion is resulting in serious stress on natural resources and on human systems, resulting in significant ecosystem breakdown and the disappearance of traditional management institutions as people are marginalised.

The general trend in the WC scenario is one of increasing water shortage that increases conflicts and restricts development and growth in most regions. Some regions however, such as the Scandinavian countries and Canada, experience no water problems as they are geographically fortunate to be located in areas of water abundance. As the date progresses to 2050, the planet is increasingly made up of closed economies which must protect their borders from environmental and economic migrants.

Most global economies are now extremely vulnerable and cannot withstand environmental shocks such as hurricanes and droughts, which cause economic collapse in some countries when they take place. Investment in long term enterprises such as technological development and science, suffer as public spending becomes limited. Investment in infrastructure is also made more difficult as more money is directed towards the military and security.

In the water sector, the incidence of water borne diseases increases significantly and more potable water is becoming polluted. Many people have lost confidence in water provision from both public and private sectors as terrorist groups threaten to poison the water supply. This leads to extreme price rises from institutions who guarantee 'safe' water. Conflict over water resources in semi arid and arid areas become more common and again there are terrorist attacks on dams and water pipelines. Private militia have set up resource enclaves of water abundance to protect supply for transnational corporations. Famines occur next to absolute shortages of water and more people die every year.

Sustainable Water World

The SWW scenario also starts in the same fashion as the CWW vision with some initial changes of direction. There is a realisation by the financial sector that investment in social institutions is essential for a sustainable future and they are willing to concede slower growth for the sake of sustainability. There is thus considerable investment in the service sector and in the environment. A significant factor in this scenario is the development of the southern economies, which experience most of the world's growth and come reasonably into line with the northern economies. This growth however, has been based on the lessons learnt by the northern countries and so environmental and social degradation are avoided.

The planet has become a global village where telecommunications result in a free movement and access to information, meaning the most remote areas feel included. Alliances are formed between NGOs, governments and the private sector (especially transnational corporations, who now have a friendly face after accepting a social charter) and there is significant emphasis on public-private partnerships for service, including water, provision. Ironically, it has been realised that absolute numbers of people were never an important global problem at a time when population growth rates begin to fall due to a fair distribution of resources.

Technology has experienced significant developments, specifically in nano-technology and telecommunications. Technology has also combined with existing local technologies (appropriate technologies) to create socially and ecologically acceptable solutions. New technologies have combined the principles of ecosystems with those of efficiency and holism.

Governance and institutional development have evolved from the lessons learnt from previous eras and the world has finally seen an end to unaccountable governments. Governance structures have become increasingly decentralised and now include a wide variety of representatives from civil society. Decision-making is now pluralistic and truly democratic and decision making structures are relied upon to resolve conflict through negotiation and consensus. This increasingly democratic trend removes much of the need for military and armed intervention. Consequently the world sees less conflict and more cohesion. When there are problems that are beyond the capacity of the nation state to deal with, for example, drought related famine, the global community responds through the United Nations which has become a mechanism for risk mitigation and minimisation.

Environmental controls are increasingly being managed by democratic global institutions which put in place regulations and management frameworks. Water was put high on the global agenda and was linked directly to development. New emphasis was put on the development of new water efficient technologies and the management of local water resources by consortiums of private sector and local social institutions. Targets were set and reached on the adequate water supply for each individual through a concerted effort by the global and national community. Water borne diseases were considerably reduced through adequate service provision and agricultural water is now efficiently used and allocated. Water withdrawals were reduced to sustainable levels and water intensity reached a remarkable historical minimum. These advances were realised through the marriage of technology, global governance and lessons learnt and use made of social water management institutions.

Future Strategies

Setting the Scene

The approach set out has emphasised the links between the management of freshwater ecosystems and the social security of communities around the world. A range of strategies that could contribute to this process. It is important to recognise that social security issues alone will not be the basis for choice of strategies. Such choices need to be based on a holistic appraisal of management needs and priorities. What is presented here is intended to stimulate a recognition of social security maintenance as one of those needs and priorities.

The goal of these strategies should be to *reduce vulnerabilities* and uncertainties through providing *greater choices* and ensuring that the choices made reflect the inherent spatial and temporal variability of both freshwater resources and patterns of human need and action. They should also work towards building a stronger societal consensus and more stable institutional environment for the future management of freshwater resources. From these departure points, such strategies should:

- *Empower local communities* to manage resources in a fair, sustainable and effective manner.
- Develop *institutional structures* that are transparent, legitimate and representative and that integrate local communities into the wider society.
- Give a *better understanding* of the role of freshwater resources in social security and of the value of these resources.
- Provide an appropriate *legal and regulatory framework* that clarifies rights and entitlements and specifies responsibilities and obligations for the maintenance of freshwater ecosystems and social security.
- Prevent or mitigate *conflicts* over freshwater resources management at all levels from the local to the international.

The general approach should be to recognise and build from what is good within existing management regimes whilst working to change or adapt the negative aspects of these regimes. In this, there is a need to think ahead, to anticipate where existing trends are moving and what the challenges of the future will be, for many of the strategies presented here are long-term in nature and will have consequences for future generations as much as for the present one. This future tense is inevitable difficult, as we are all aware of the fragility of predictions and projections. What it does mean is that any such strategies should be *flexible* and should have the capacity to *cope with uncertainty*.

One of three commissioned by IUCN to provide building blocks for the overall development of a "vision" on water and nature. To ensure that there is a level of coherence between the three papers, where possible the strategies identified here are divided into:

- *Adaptation Strategies*: strategies that recognise that it is either not possible or not desirable to remove a specified source of vulnerability and consequently seek to enhance capacities to cope with it. An example is the threat posed by extreme events such as cyclones or trends that are beyond immediate control such as global warming.
- *Mitigation Strategies*: strategies that are intended to significantly reduce or eliminate defined threats or sources of degradation: for example,

the removal of a particular point of pollution or new legislation to clarify rights over resources.

There are, however, key aspects of the challenge that are not directly connected to individual types of vulnerability and consequently do not fit easily into this division between adaptation and mitigation. These are issues related to the *structural conditions of society* within which the management of freshwater ecosystems and the reproduction of social security occur. In particular, these reflect the structure of institutions within which different forms of interaction take place and the underlying social conditions within which these institutions operate. In other words, these define the context within which individual strategies, whether for adaptation or mitigation, will be placed.

Strategies for Structural Change

These strategies are intended to change the *societal and institutional context* within which threats to social security and freshwater ecosystems are addressed, whether through adaptation or mitigation approaches. The emphasis is, inevitably, on policy and institutional issues, with in particular an emphasis on moving towards the greater horizontal and vertical integration. Actions are needed at all levels, from the local to the global, but there is a particular need to look carefully at the structure and operation of the state and the links between state agencies and the wider civil society. The key areas of such actions are:

International Co-operation

The origins of many threats to ecosystems and social security are beyond the borders of many individual states. They can be global: for example, the consequences of climate change. They can represent hydrological dynamics within a river basin that crosses national borders: upstream pollution or over-abstraction. Addressing these threats entails co-operation between countries, not least to try to remove the threat of overt conflict. We have seen that there are positive examples of co-operation (such as the Bangladesh-India agreement on the water of the Ganges) but even these tend to be piecemeal and there is rarely a perspective that recognises the overall dynamics of such threats (e.g. a comprehensive Ganges-Brahmaputra Basin agreement).

What does exist is an emerging international consensus on the direction and goals of freshwater ecosystems management. This consensus emphasises the importance of holistic approaches that define rights and responsibilities and that integrate the preservation of ecosystem functions as a central goal of water resources policies. Major international initiatives, such as the work of the Commission for Sustainable Development and the GEF Operational Strategy, provide an opportunity for articulating the relationship between social security and the management of water resources with the goal of building a clearer international consensus and influencing the policies and approaches of individual

This consensus needs to be built on through the following strategies:

- The development of *co-ordinated strategies* to cope with threats such as global warming, major storms, floods and droughts that affect the integrity of ecosystems and the security of people. This should include joint risk assessments, collaboration on early warning and response systems and mutual aid when disasters strike.
- The establishment of *river basin forums* that integrate all countries (including governments and representations from civil society) within major river basins where there are threats to ecosystem integrity or social security. Such forums will rarely have formal powers, but should be a basis for reaching better understanding and establishing consensus on the future of the river basin. They can also play a crucial role in the defusing of conflicts over the sharing of water resources along the river basin.
- Creating a *better understanding* of all aspects of freshwater resources and their uses within river basins and across international borders through shared research, information and monitoring systems. This should include both the ecological dynamics and the full valuation (including social security values) of the resources and their uses.

Policy Reforms

Many aspects of vulnerability and the degradation of freshwater ecosystems are rooted in incomplete or inappropriate legal and policy frameworks at the national level. Above all, it relates to who has control, with many countries having a long history of the state expropriating rights over freshwater ecosystems and then failing to use these powers in an effective manner. This need for a range of policy reforms should centre on:

- *Water and Land Tenure Laws*: providing a legal basis for assigning rights and entitlements (including customary and communal rights to common property resources) is crucial for the maintenance of both ecosystems and social security.

- *Policy Priorities* in food production, industrialisation and water resources allocation are all critical. Past policies that have given primacy to high-intensity grain production and/or industrialisation with no regard to resource sustainability have been particularly damaging. All policies need to take full account of their wider impacts on people and ecosystems.
- *Investment and Budgetary Priorities*: the state will continue to play a key role in many parts of the world. The allocation of resources to ecosystems maintenance and meeting the needs of vulnerable groups is critical in establishing coping strategies.
- The establishment and enforcement of *appropriate standards* (both quality and quantity) for the use of water resources, based on the dual goals of enhancing social security and conserving freshwater ecosystems.
- Defining and implementing meaningful *pricing or cost recovery* mechanisms that reflect the true value of the resources and that ensure 'polluter pays' principles are implemented.

Institutional Restructuring

The need for more accountable, effective and devolved institutions has already been discussed. There are trends towards this in many parts of the world. These need to be encouraged through the demonstration of the importance for these trends for the maintenance of both social security and ecosystems integrity. The specifics of such strategies will vary from place to place, but general trends should include:

- *Improving Efficiency*: too many state agencies, in particular, are centralised and bureaucratic in nature, and operate to prescribed rules and procedures with little thought as to why things are done the way they are. There is a move in many organisations (including major multinationals) towards management structures that devolve authority to the operational level and operate through shared value systems, cultures and trust. These principles need to be inculcated into government agencies that influence the ways in which freshwater ecosystems are managed.

- *Multi-Agency Subsidiarity:* there is a need for multiple agency involvement, including central government departments, local government, NGOs and locally-based community organisations in the management of freshwater resources. The division of responsibilities between these different levels and workable protocols for their interaction and co ordination are similarly essential. This is an inherently complex process which involves re-defining mandates and levels of authority, re-directing budgets, capabilities and facilities and establishing a clear and shared vision on where water resources management is going. This type of restructuring will bring benefits in both more effective operations (and lower costs) and a clearer social consensus on what should be done and why.
- *Linking to Democratisation and Decentralisation:* the social consensus should also be part of a wider process of development and change. Many parts of the world are seeing strengthening democratic processes and the decentralisation of state structures. Both are desirable trends, as both involve the wider civil society in decision-making and produce a stronger link between the local level (where resources and managed and social security operates) and external agencies. These links are also essential in providing the mechanisms for the arbitration of potential conflicts within or between different communities over the use and preservation of freshwater ecosystems.
- *Strengthening Local Institutions:* the examples cited above have demonstrated the potential of local institutions in water resources management, but it is important not to idealise these initiatives. Although they are well-developed in some areas, they are weak in others and will need considerable support if they are to become the focal point of local- level management of water resources and infrastructure. This local-level capacity- building is the basis upon which many strategies should be based, for without it there is no hope of implementing reforms that rely on the active involvement of local groups.
- *Improving Understanding and Planning Capabilities*: knowing what to do and how to do it is the key to sustainable changes. This is in turn dependent on good information about the present situation and effective tools to analyse and predict the consequences of change. There is often a dearth of both, with limited and technically-oriented data and

planning tools that are rarely widely accessible. There is a clear need to develop information bases and analytical tools that provide are accessible to all and that provide a structure within which social consensus on what to do can be based.

Adaptation Strategies

These strategies centre on the development or enhancement of coping mechanisms where people and communities can improve their abilities to live with forces that they cannot do anything about. The development of improved capacities to cope with existing trends is a critical issue. In some cases, this will build on what is already there (and this includes supporting existing coping strategies that are being undermined). In others, it will entail a whole new set of responses to threats that are perhaps new themselves. In both cases, the overall objective is to create the circumstances where society as a whole works to support those sections of the community (and those ecosystems) that are most at risk from these threats.

In this, the emphasis is on the empowerment of local communities: the creation of circumstances where they have greater control over their own lives and greater choices about how they will respond to the challenges they face. This does not mean, however, that all actions are aimed solely at the local level, for there are a range if actions needed at all levels from the individual to the international and the success of the adaptation (or mitigation) strategies identified here will often be contingent on the structural changes outlined above.

The main focus of such adaptation strategies will consequently be to improve *community capabilities to manage freshwater ecosystems*, and especially common property resources, that are experiencing change caused by forces beyond control (such as climate or economic change). The direction of these improvements will be to maximise the sustainable productivity of these ecosystems, but to do so in ways that does not threaten their long-term integrity. The details will depend on local conditions (economic and ecological), but the goal should be to recognise and find the balance between all potential goods and services these ecosystems offer. Within this broad field, some specific strategies are:

- Ensuring that the community have clear *rights and entitlements*, supported by the jurisdiction of appropriate external agencies.

- Working with local communities to *adapt traditional harvesting* of water or water- based plants and animals to reflect changing sustainable potentials.
- *Identifying new potentials*, including non-consumptive uses such as ecotourism, to compensate for declining resource utility from traditional forms of exploitation.
- Developing new *technical options*, including management regimes and technology choices, that conserve the resources more effectively.
- Identifying *alternative sources* to meet existing needs, such as replacing wild gathered fish with aquaculture or using groundwater instead of surface water for domestic needs.
- Assisting local communities to develop *disaster response* capabilities where environmental threats from storms, floods, droughts and so on are more severe. This should include both organisational issues and structural works such as re-establishing mangroves or constructing refuges.

Mitigation Strategies

Mitigation strategies are those approaches that are aimed at lessening or removing the effects of threats to freshwater ecosystems that undermine social security. In these instances, the basic assumption is that something can and should be done to deal with immediate threats. They are typically more amenable to local control, but this does not mean that they are purely localised issues: they are often as dependent upon the wider context as the adaptation issues discussed above. It does mean that the central thrust of these strategies is to enhance the capability of local communities and the wider society to reverse many of the threats to freshwater ecosystems found in different parts of the world today. Again, in most cases there will need to be a combination of social/institutional and technical/management interventions. Examples of such mitigation strategies are:

- Ensuring that the community have clear *rights and entitlements* (always a pre-requisite), supported by the jurisdiction of appropriate external agencies, as ecosystem degradation is often linked to inappropriate or uncertain rights.
- Creating effective *local institutional capacities* to provide a vehicle for empowering local communities and establishing sustainable freshwater ecosystems management.

- Creating mechanisms for *conflict mitigation* between local communities and outside agencies, between different local communities and between groups within a locality. This is often essential for reversing unsustainable practices that competition creates.
- Identifying and implementing *sustainable multiple use management strategies* that are based on reversing degrading practices whilst retaining the benefits that the resource uses bring.
- Developing a better *understanding* of both the origins of present problems and the options for change, with a key challenge being to ensure that such knowledge development involves all stakeholders and integrated both *indigenous and external knowledge systems*.
- Defining appropriate *interventions*, including *investments* that are both technically sound and economically wise. This will often require some level of external resources, but should ensure that there is an involvement of local resources as well.
- Developing appropriate *resource charges and cost recovery mechanisms*, to ensure that the users of freshwater resources pay for the benefits they receive. Such charges should be linked to local abilities to pay, with local communities receiving the benefits of such cost recovery.
- There is a clear need to create *a political consensus* for many of the actions needed to mitigate unsustainable ecosystems management. This in turn means that there is a need for effective *advocacy* within civil society to provide an understanding of the implications of present paths and the need for change that may entail short-term costs to ensure long-term benefits.

Creating a Vision for the Future

It is hoped and intended that the reader will consider each of the points raised below in relation to her/his life: both (for many) as a professional whose working existence is based around dealing with these issues and as a citizen, a member of a particular society and the global community whose future is as dependent on these issues as that of the rest of the planet.

The Vision

We all face an uncertain future, surrounded by fears but supported by hopes and dreams. What should the future of water be, for you and for the community in which you live? Where would you hope that your society and the global community will be in the next century and how can the management of freshwater ecosystems contribute to this through reducing uncertainties and providing greater security.

The Challenges

To manage freshwater ecosystems in the 21st Century in ways that are wise, that recognise the inter-dependence of different parts of the hydrological cycle and the needs of different sections of the community. In defining our goals for the management of these ecosystems, how far can we incorporate their role in providing greater security to livelihoods? Can we make the best use of all their potentials in ways that do not undermine their integrity? Can we bring together people, communities and nations to reach a consensus on the distribution of these benefits without the conflicts that threaten to be the source of insecurity and suffering in the 21st Century?

The World Today

All around us we see worrying trends, as freshwater resources are degraded by careless use and abuse, more conflicts emerge and the poor access to these resources means that too many millions still suffer from illness, poverty, drudgery and insecurity. Amid this gloom, however, there are signs of hope, in the ingenuity and creativity of many communities in working together to overcome the challenges they face and the better understanding of these human potentials that is now emerging. Can we learn from today, both to tell us what not to do in the future and to find the experiences and examples that form the basis of hope for the future? More importantly, can we help inform others of these hopes and create the social consensus for change that is so important if the signs of despair are to be overcome?

Creating Understanding

These issues are far from simple. Too often a real consensus is not possible because people start from different assumptions and analyse the problems with different reference frames. Can we find a conceptual framework that helps inform us and helps us to inform others?

What Does The Future Hold?

The future is always an uncertain place, but we can identify trends that give us direction and provide insights into the consequences of certain courses of action (or inaction). To borrow from Sergio Leone, we can see The Good (a transition to sustainable freshwater ecosystems management where their contribution to social security is conserved and enhanced), The Bad (a conventional world, with 'business-as-usual' meaning a gradual deterioration of the resources and declining social security) and The Ugly (a water crisis world in which ever more demand chases ever declining resources and oppression, conflict and despair is the lot of us all).

What Can Be Done?

What is clear is that these challenges need vigorous and sustained actions now if they are to be addressed. The key, in the context of social security and freshwater ecosystems management, is to create the social and institutional context within which sustainable management and the empowerment of the marginal is possible. This means different things in different places. What should (and can) be done first? How can the power of entrenched interests be overcome? How can we ensure that we really reach and empower the most in need to take control over their own lives and have greater choices to overcome the challenges they face?

References

Agnew. C. and Anderson. E. (1992) *Water resources in the arid realm* Routledge, London.

Goswami. A. (1998) *Troubled waters* Business India. August 10-23: 62-68.

McCafferey. S.C. *Water politics and international law* in Gleick. P.H. (Ed.) Water in crisis: a guide to the World's freshwater resources. Oxford University Press. London.

McDonald, A.T. and Kay, D. (1986) *Water Resources Issues and Strategies*. Longman, London.

Wallensteen. P. and Swain. A. (1996) *International freshwater resources: source of new conflict*. Department of Peace and Conflict Research. Uppsala University. Sweden.

World Bank (1992) *World Development Report 1992*, Development and Environment, New York: Oxford University Press.

World Resources Institute (1998) *A Guide to the Global Environment* Oxford University Press, Oxford UK.

4

Ideas to Maintain the Integrity of Freshwater Ecosystems

Freshwater is vital to human life and economic well-being, and societies draw heavily on rivers, lakes, wetlands, and underground aquifers to supply water for drinking, irrigating crops, and running industrial processes. The benefits of these extractive uses of freshwater have traditionally overshadowed the equally vital benefits of water that remains in stream to sustain healthy aquatic ecosystems. There is growing recognition that functionally intact and biologically complex freshwater ecosystems provide many economically valuable commodities and services to society. The services supplied by freshwater ecosystems include flood control, transportation, recreation, purification of human and industrial wastes, habitat for plants and animals, and production of fish and other foods and marketable goods. These human benefits are what ecologists call ecological services, defined as "the conditions and processes through which natural ecosystems, and the species that make them up, sustain and fulfill human life."

Sustaining Healthy Freshwater Ecosystems

Over the long term, healthy freshwater ecosystems are likely to retain the adaptive capacity to sustain production of these ecological services in the face of future environmental disruptions such as climate change. Ecological services are costly and often impossible to replace when aquatic ecosystems are degraded. Yet today, aquatic ecosystems are being severely altered or destroyed at a greater rate than at any other time in human history, and far faster than they are being restored. Debates involving sustainable allocation of water resources should recognize that maintenance of freshwater ecosystem integrity is a legitimate goal that must be considered among the competing demands for freshwater. Coherent policies are required that more equitably allocate water resources between natural ecosystem functioning and society's extractive needs.

Current water management policies in the United States are clearly unable to meet this goal. Literally dozens of different government entities have a say in what wastes can be discharged into water or how water is used and redistributed, and the goals of one agency are often at cross-purposes with those of others. U. S. laws and regulations concerning water are implemented in a management context that focuses primarily on maintaining the lowest acceptable water quality and minimal flows, and protecting single species rather than aquatic communities. A fundamental change in water management policies is needed, one that embraces a much broader view of the dynamic nature of freshwater resources and the short- and long-term benefits they provide.

Our current educational practices are as inadequate as management policies to the challenge of sustainable water resource management. Hydrologists, engineers, and water managers, the people who design and manage the nation's water resource systems, are rarely taught about the ecological consequences of management policies. Likewise, ecologists are rarely trained to consider the critical role of water in human society or to understand the institutions that manage water. Economists, developers, and politicians seldom project far enough into the future to fully account for the potential ecological costs of short-term plans. Few Americans are aware of the infrastructure that brings them pure tap water or carries their wastes away, and fewer still understand the ecological tradeoffs that are made to allow these conveniences. Although the requirements of healthy freshwater ecosystems are often at odds with human activity, this conflict need not be inevitable.

The challenge is to determine how society can extract the water resources it needs while protecting the important natural complexity and adaptive capacity of freshwater ecosystems. Current scientific understanding makes it possible to outline here in general terms the requirements for adequate quantity, quality, and timing of water flow to sustain the functioning of freshwater ecosystems. A critical next step will be communication of these requirements to a broader community. The American public, when given information about management alternatives, supports ecologically based management approaches, particularly toward freshwater.

Table 1. Changes in hydrologic flow, water quality, wetland area, and species viability in U.S. rivers, lakes, and wetlands since Euro-American settlement.

U. S. Freshwater Resources	*Pre-settlement Condition*	*Current Conditions*	*Source*
Undammed rivers (in 48 contiguous states)	5.1 million km	4.7 million km	Echeverria et al. 1989
Free-flowing rivers that qualify for wild and scenic status (in 48 contiguous states)	5.1 million km	0.0001 million km	US DOI 1982
Number of dams >2m	0	75,000	CEQ 1995
Volume of water diverted from surface waters	0	10 million m^3 day^{-1}(1985)	Solley et al. 1998
Total daily U. S. water use	Unknown	1.5 million m^3 day^{-1}(1995)	Solley et al. 1998
Sediment inputs to reservoirs	not applicable	1,200 million m^3/year	Stallard 1998
River water quality* (1.1 million km surveyed)	Unimpaired	402,000 km impaired*	EPA 1998
Lake water quality* (6.8 million ha surveyed)	Unimpaired	2.7 million ha impaired*	EPA 1998
Wetland acreage (in 48 contiguous states)	87 million ha	35 million ha	van der Leeden et al. 1990
Number of native fresh-water fish species	822 species	202 imperiled or extinct	Stein and Flack 1997
Number of native fresh-water mussel species	305 species	157 imperiled or extinct	Stein and Flack 1997
Number of native crayfish species	330 species	111 imperiled or extinct	Stein and Flack 1997
Number of native amphibian species	242 species	64 imperiled or extinct	Stein and Flack 1997

Several previous studies that have addressed the overall condition of freshwater resources have recognized that

- water movement through the biosphere is highly altered by human activities;

- water is intensively used by humans;
- poor water quality is pervasive;
- and freshwater plant and animal species are at greater risk of extinction from human activities compared with all other species. These and other analyses indicate that freshwater ecosystems are under stress and at risk.

Clearly, new management approaches are needed. We suggest steps to be taken toward restoration and conclude with recommendations for protecting and maintaining freshwater ecosystems.

Requirements For Freshwater Ecosystem Integrity

Freshwater ecosystems differ greatly from one another depending on type, location, and climate, but they nevertheless share important features. For one, lakes, wetlands, rivers, and their connected ground waters share a common need for water within a certain range of quantity and quality. In addition, because freshwater ecosystems are dynamic, all require a range of natural variation or disturbance to maintain viability or resilience. Water flows that vary both season to season and year to year, for example, are needed to support plant and animal communities and maintain natural habitat dynamics that support production and survival of species. Variability in the timing and rate of water flow strongly influence the sizes of native plant and animal populations and their age structures, the presence of rare or highly specialized species, the interactions of species with each other and with their environments, and many ecosystem processes. Periodic and episodic water flow patterns also influence water quality, physical habitat conditions and connections, and energy sources in aquatic ecosystems. Freshwater ecosystems, therefore, have evolved to the rhythms of natural hydrologic variability.

The structure and functioning of freshwater ecosystems are also tightly linked to the watersheds, or catchments, of which they are a part. Water flowing through the landscape on its way to the sea moves in three dimensions, linking upstream to downstream, stream channels to floodplains and riparian wetlands, and surface waters to ground water. Materials generated across the landscape ultimately make their way into rivers, lakes, and other freshwater ecosystems. Thus these systems are greatly influenced by what happens on the land, including human activities.

We have identified five dynamic environmental factors that regulate much of the structure and functioning of any aquatic ecosystem, although their relative importance varies among aquatic ecosystem types. The interaction of these drivers in space and time defines the dynamic nature of freshwater ecosystems:

1. The *flow pattern* defines the rates and pathways by which rainfall and snowmelt enter and circulate within river channels, lakes, wetlands, and connecting ground waters, and also determines how long water is stored in these ecosystems.
2. *Sediment and organic matter inputs* provide raw materials that create physical habitat structure, refugia, substrates, and spawning grounds and supply and store nutrients that sustain aquatic plants and animals.
3. *Temperature and light characteristics* regulate the metabolic processes, activity levels, and productivity of aquatic organisms.
4. *Chemical and nutrient conditions* regulate pH, plant and animal productivity, and water quality.
5. The *plant and animal assemblage* influences ecosystem process rates and community structure.

In naturally functioning freshwater ecosystems, all five of these factors vary within defined ranges throughout the year, tracking seasonal changes in climate and day length. Species have evolved and ecosystems have adjusted to accommodate these annual cycles. They have also developed strategies for surviving – and often requiring — periodic hydrologic extremes caused by floods and droughts that exceed the normal annual highs or lows in flows, temperature, and other factors.

Focusing on one factor at a time will not yield a true picture of ecosystem functioning. Evaluating freshwater ecosystem integrity requires that all five of these dynamic environmental factors be integrated and considered jointly.

Flow Conditions for Rivers and Streams

Base flow conditions characterize periods of low flow between storms. They define the minimum quantity of water in the channel, which directly influences habitat availability for aquatic organisms as well as the depth to saturated soil for riparian species. The magnitude and duration of base flow varies greatly among different rivers, reflecting differences in climate, geology, and vegetation in a watershed.Frequent (that is, two-year return interval) floods reset the system by flushing fine materials from the streambed, thus promoting higher production during base flow periods. High flows may also facilitate dispersal of organisms both up- and downstream. In many cases moderately high flows inundate adjacent floodplains and maintain riparian vegetation dynamics.

Rare or extreme events such as 50- or 100-year floods represent important reformative events for river systems. They transport large amounts of sediment, often transferring it from the main channel to floodplains. Habitat diversity within the river is increased as channels are scoured and reformed and successional dynamics in riparian communities and floodplain wetlands are reset. Large flows can also remove species that are poorly adapted to dynamic river environments such as upland tree species or non-native fish species. The success of non-native invaders is often minimized by natural high flows, and the restriction of major floods by reservoirs plays an important role in the establishment and proliferation of exotic species in many river systems.

Seasonal timing of flows, especially high flows, is critical for maintaining many native species whose reproductive strategies are tied to such flows. For example, some fish use high flows to initiate spawning runs. Along western rivers, cottonwood trees release seeds during peak snowmelt to maximize the opportunity for seedling establishment on floodplains. Changing the seasonal timing of flows has severe negative consequences for aquatic and riparian communities.

Annual variation in flow is an important factor influencing river systems. For example, year-to-year variation in runoff volume can maintain high species diversity. Similarly, ecosystem productivity and foodweb structure can fluctuate in response to this year-to-year variation. This variation also ensures that various species benefit in different years, thus promoting high biological diversity.

Flow Patterns

An evaluation of the characteristics required for healthy functioning can begin with a description of the natural or historical flow patterns for streams, rivers, wetlands and lakes. Certain aspects of these patterns are critical for regulating biological productivity (that is, the growth of algae or phytoplankton that form the base of aquatic food webs) and biological diversity, particularly for rivers. These aspects include base flow, annual or frequent floods, rare and extreme flood events, seasonality of flows, and annual variability. Such factors are also relevant for evaluating the integrity of lakes and wetlands because flow patterns and hydroperiod (that is, seasonal fluctuations in water levels) influence water circulation patterns and renewal rates, as well as types and abundances of aquatic vegetation such as reeds, grasses, and flowering plants. Furthermore, the characteristic flow pattern of a lake, wetland, or stream critically influences algal productivity and is an important factor to be considered when determining acceptable levels of nutrient (nitrogen and phosphorus) runoff from the surrounding landscape.

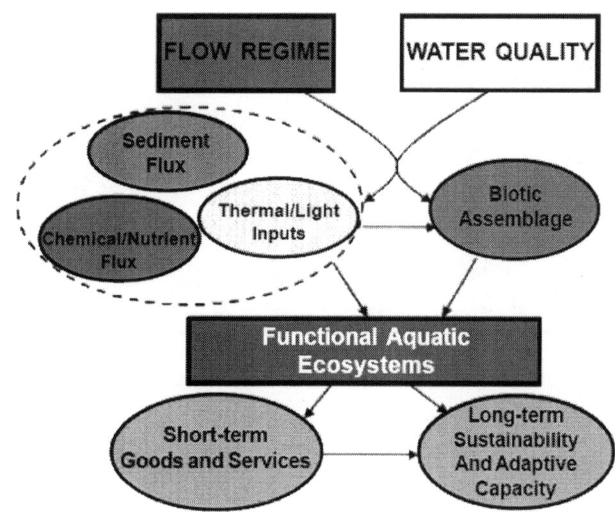

Figure 1. Conceptual model of major forces that influence freshwater ecosystems.

Human alterations of river flow have seldom taken into account the ecological consequences. "Many rivers now resemble elaborate plumbing works, with the timing and amount of flow completely controlled, like water from a faucet, so as to maximize the rivers' benefits for humans," wrote water policy expert Sandra L. Postel. "But while modern engineering has been remarkably successful at getting water to people and farms when and where they need it, it has failed to protect the fundamental ecological function of rivers and aquatic systems."

Rivers in the U.S. West are prime examples of how human manipulation of water flows can lead to multiple damages to riverbank and floodplain processes and communities. Damming rivers and dampening natural variations in flow rates by maintaining minimum flows year round have contributed to widespread loss of native fish species and regeneration failure of native cottonwood trees, which used to support diverse riparian communities.

Sediment and Organic Matter Inputs

In river systems, the movement of sediments and influxes of organic matter are important components of habitat structure and dynamics. Natural organic matter inputs include seasonal runoff and debris such as leaves and decaying plant material from land-based communities in the watershed. Especially in smaller rivers and streams, the organic matter that arrives from the land is a particularly important source of energy and nutrients, and tree trunks and other woody materials that fall into the water provide important substrates and habitats for aquatic organisms. Natural sediment movements are those that accompany natural variations in water flows. In lakes and wetlands, all but the finest inflowing sediment falls permanently to the bottom, so that over time these systems fill. The invertebrates, algae, bryophytes, vascular plants, and bacteria that populate the bottoms of freshwater systems are highly adapted to the specific sediment and organic matter conditions of their environment, as are many fish species, and do not persist if changes in the type, size, or frequency of sediment inputs occur. The fate of these organisms is critical to sustaining freshwater ecosystems since they are responsible for much of the work of water purification, decomposition, and nutrient cycling.

Humans have severely altered the natural rates of sediment and organic matter supply to aquatic systems, increasing some inputs while decreasing others. Poor agricultural, logging, or construction practices, for example, promote high rates of soil erosion. In many areas small streams or wetlands have even been completely eliminated through filling, paving, or re-routing into artificial channels. The U.S. Environmental Protection Agency (EPA) reports that in one quarter of all lakes with sub-standard water quality, the cause of impairment is silt entering from agricultural, urban, construction, and other non-point (widely dispersed) sources. Dams alter sediment flows both for the reservoirs behind them and the streams below, silting up the former while starving the latter. By one estimate, another 1.2 billion cubic meters of sediment builds up each year in U. S. reservoirs. This sediment capture in turn cuts off normal sand, silt, and gravel supplies to downstream reaches, causing streambed erosion that both degrades in-channel habitat and isolates floodplain and riparian wetlands from the channel during rejuvenating high flows. Channel straightening, overgrazing of river and stream banks, and clearing of streamside vegetation reduce organic matter inputs and often increase erosion.

Temperature and Light

The light and heat properties of a body of water are influenced by climate and topography as well as by the characteristics of the water body itself: its chemical composition, suspended sediments, and algal productivity. Water temperature directly regulates oxygen concentrations, the metabolic rate of aquatic organisms, and associated life processes such as growth, maturation, and reproduction. The temperature cycle greatly influences the fitness of aquatic plants and animals and, by extension, where species are distributed in the system and how the living community in a body of water varies from season to season. In lakes particularly, the absorption of solar energy and its dissipation as heat are critical to development of temperature gradients between the surface and deeper water layers and also to water circulation patterns. Circulation patterns and temperature gradients in turn influence nutrient cycling, distribution of dissolved oxygen, and both the distribution and behavior of organisms, including game fishes. Water temperature can change dramatically downstream of dams. In Utah's Green River, mean monthly water temperatures ranged between 2 degrees Celsius (C) in winter and 18 degrees C in summer before completion of the Flaming Gorge Dam in 1962. After dam closure, the annual range of mean monthly water temperatures below the dam was greatly narrowed, to between 4 C and 9 C. As a result, species richness declined and 18 genera (that is, groups of related species) of insects were lost; other species, notably freshwater shrimp, came to dominate the ranks of invertebrate animals. Aquatic insects have not recovered despite 20 years of partial temperature restoration achieved by releasing water from warmer reservoir water layers. Water temperature also dropped in the Colorado River after closure of the Glen Canyon Dam in 1963, and there was a dramatic increase in water clarity. Water clarity now routinely allows visibility to greater than 7 meters, whereas prior to dam closure, the water column was opaque with suspended sediments. The colder, clearer waters have allowed a non-native trout population to flourish, at the top of an unusual food web more commonly found much further north.

Nutrient and Chemical Conditions

Natural nutrient and chemical conditions are those that reflect local climate, bedrock, soil, vegetation type, and topography. Natural water conditions can range from clear, nutrient-poor rivers and lakes on crystalline bedrock to much more chemically enriched and algae-producing freshwaters in catchments with organic matter-rich soils or limestone bedrock. This natural regional diversity in watershed characteristics, in turn, sustains high biodiversity.

A condition known as cultural eutrophication occurs when additional nutrients, chiefly nitrogen and phosphorus, from human activities enter freshwater ecosystems. The result is a decrease in biodiversity, although productivity of certain algal species can increase well beyond original levels. Midwestern and Eastern lakes such as Lakes Michigan, Huron, Erie, and Ontario demonstrate the consequences of excess inputs of nutrients and toxic contaminants, as well as non-native species introductions and over-fishing. Onondaga Lake, New York, which was polluted with salt brine effluent from a soda ash industry, likewise responded with marked changes in the plankton and fish communities, including invasions by non-native fish species. Among U.S. lakes identified by the EPA as impaired in 1996, excess nutrients contributed to more than half of the water quality problems. More than half of agricultural and urban streams sampled by the U. S. Geological Survey were found to have pesticide concentrations that exceed guidelines for the protection of aquatic life.

Plant and Animal Assemblages

The community of species that lives in any given aquatic ecosystem reflects both the pool of species available in the region and the abilities of individual species to colonize and survive in that water body. The suitability of a freshwater ecosystem for any particular species is dictated by the environmental conditions – that is, water flow, sediment, temperature, light, and nutrient patterns — and the presence of, and interactions among, other species in the system. Thus, both the habitat and the biotic community provide controls and feedbacks that maintain a diverse range of species. The high degree of natural variation in environmental conditions in freshwaters across the United States promotes high biological diversity. In fact, North American freshwater habitats are virtually unrivaled in diversity of fish, mussel, crayfish, amphibian, and aquatic reptile species compared with anywhere else in the world. The biota, in turn, are involved in shaping the critical ecological processes of primary production, decomposition, and nutrient cycling. Within a body of water, species often perform overlapping, apparently redundant roles in these processes, a factor that helps provide local ecosystems with a greater capacity to adapt to future environmental variation. High apparent redundancy (that is, species richness or biodiversity) affords a kind of insurance that ecological functions will continue during environmental stress. Critical to this is connectivity among water bodies, which allows species to move to more suitable habitat as environmental conditions change.

Human activities that alter freshwater environmental conditions can greatly change both the identity of the species in the community and the functioning of the ecosystem. Excessive stress or simplification of natural complexity has the potential to push functionally intact freshwater ecosystems beyond the bounds of resilience or sustainability, threatening their ability to provide important goods and services on both short and long time scales. Further, introduction of non-native species that can thrive under the existing or altered range of environmental variation can contribute to the extinction of native species, severely modify food webs, and alter ecological processes such as nutrient cycling. Exotic species are often successful in modified systems, where they can be difficult to eradicate.

Tools Available for Restoration

Despite widespread degradation of freshwater ecosystems, management techniques are available that can restore these systems to a more natural and sustainable state and prevent continued loss of biodiversity, ecosystem functioning, and ecological integrity. One technique, for example, involves restoring some of the natural variations in stream flow, based on the understanding that river systems are naturally dynamic. New statistical approaches to setting management targets for streamflow variability over time have been applied to or proposed for several rivers, including the Flathead River in Montana, the Roanoke River in North Carolina, and the vast Colorado River system in the West. These variable streamflow techniques seek a balance between water delivery needs for power generation or irrigation, and in-stream ecological needs for flow variability that displays a certain timing, frequency, duration, and rate of change characteristic of the natural syste. Restoring this flow variability helps to reconnect dynamic riparian and groundwater systems with surface flows, enabling water to move more naturally through all the spatial dimensions that are essential to fully functional ecosystems. Other restoration efforts target pollution, both from point sources such as effluent from industrial or sewage pipes and nonpoint sources such as fertilizer runoff from urban lawns and rural croplands. Point sources of water pollution are readily identified, and many have been controlled, thanks in large part to the federal Clean Water Act and Safe Drinking Water Act.

Nonpoint sources of nutrients and toxins now supply the majority of pollutants to freshwater ecosystems. In some situations, best management practices have succeeded in reducing runoff of agricultural pollutants. These practices include erosion control and moderate applications of fertilizers, pesticides and herbicides. Best management practices require willing farmers, however, and willingness is often a response either to economic incentives or to stringentregulations. To help in determining best management practices, the EPA has recently published guidelines for establishing acceptable nutrient runoff criteria for different regions of the United States, recognizing the inherent natural variability in local and regional availability of nutrients.

The guidelines are based on Total Maximum Daily Load (TMDL), a calculation of the maximum amount of a pollutant that a water body can receive and still meet water quality standards. To allow for natural variation, water quality standards for a pollutant are established within each ecoregion based on comparison with relatively unpolluted waters or – if few or no unpolluted waters remain in a region — on waters with the lowest pollution levels. Once a standard is set, management practices can be enacted to reduce inputs of unwanted pollutants.

Another large source of nonpoint pollution is atmospheric deposition of nitrogen and other contaminants that fall as acid rain or dry pollutants. These could be reduced through more stringent controls on emissions of sulfur, nitrogen, metals, and organic toxins, and through development and application of more efficient transportation and energy production technologies.

Study of The Colorado River

The Colorado River is one of the most highly regulated and heavily used river systems in the world. Two principal reservoirs, Lakes Powell and Mead, along with 12 other large reservoirs store and release water according to complicated equations designed to maximize both hydroelectric generation and water supplies for agricultural, domestic, and industrial use in seven states across the Western United States and Mexico. More than 30 million people depend on Colorado River water. The original Colorado River Compact of 1928 allocated all water for societal use. (Actually it over-allocated because typical water volumes were overestimated while year-to-year variability was ignored.)

Physical changes to the river below the dams have been profound. Flow in the Colorado River is snowmelt driven, and pre-dam flow patterns were dominated by large discharges from April through July, followed by low flows in late summer and fall. The river carried tremendous amounts of sediment from the highly erodible Colorado Plateau, and river temperatures were seasonally warm. Today, river flow is nearly decoupled from natural snowmelt, and peak discharges can occur in any month, often November to January. Daily changes in water releases as great as 566 cubic meters per second occur regularly for hydropower generation. Alluvial sediment, which once played a vital role in creating in-channel habitat, is now trapped behind the dams, and the waters below are clear and sediment-starved. Also, because water is released from the bottom waters of most reservoirs, water temperatures for hundreds of kilometers below the dams are very cold throughout the summer and relatively warm during the winter, a reversal of the natural seasonal cycle. (An exception is Flaming Gorge Reservoir on the Green River in the upper Colorado basin, where water is released from multiple reservoir layers.)

Ecological responses to the dams have been equally profound. The clear, cold tail waters below the dams, in conjunction with widespread introduction of non-native species, have promoted food webs that are alien to the Colorado River. Prior to regulation, the organic matter that fueled the river food web primarily originated on land and was carried into the river during large runoff events. Now, organic matter is supplied largely by luxuriant mats of algae that grow on the bottom of the river. The algae are consumed by insects and other invertebrates that historically occurred only in the much colder tributaries of the Colorado; these insects and invertebrates are in turn eaten by non-native rainbow and brown trout. Below the Glen Canyon Dam that holds Lake Powell, only four out of eight indigenous fish species remain, along with 22 non-native fishes, many of which either compete with or directly feed on the endangered native fish. Native cottonwood trees and the animal community they support are declining because the trees are unable to take root under variable flows. Also, upstream reservoirs that reduce the magnitude of annual floods prevent the establishment of cottonwoods higher on the riverbanks. Other shrubs and trees that are more tolerant of these modified conditions have grown profusely, including non-natives such as tamarisk.

The effects of 14 major dams and hundreds of water diversions have been felt all the way to the river mouth. Since completion of the Glen Canyon Dam in 1963, measurable flows from the Colorado River into the Sea of Cortez have occurred only infrequently. The wetland area at the mouth of the river has decreased from a historical average of 250,000 hectares to 5,800 to 63,000 hectares (depending on the year). In the Sea of Cortez, the lack of freshwater inflows has contributed to the endangerment of a large number of species, and the loss of algal productivity has caused the abundance of bivalve mollusk populations to drop 94 percent from 1950 values.

To reduce the impact of dam operations on the river's ecological resources, Congress passed the Grand Canyon Protection Act of 1992. A large group of Colorado River stakeholders now work with a Department of Interior sponsored Grand Canyon Monitoring and Research Center to attempt through adaptive management to protect and restore riparian areas and native fishes, several of which are threatened or endangered. In 1996, after nearly 15 years of study, a large experimental flood was generated to help scientists and managers investigate the effects of high flows on sediment transport and biological, cultural, and socioeconomic resources. Another set of experimental floods is planned, along with aggressive efforts to reduce non-native trout populations. There is also discussion of installing a thermal control device on Glen Canyon Dam to raise water temperatures below it. Partial restoration of historic temperatures below Flaming Gorge Dam on the Green River, however, have not improved conditions for aquatic insects directly below the dam. More than 20 years later, the number of species is as low or lower than before the restoration efforts began. Further downstream, the number of insect taxa did increase, but only because warmer summer temperatures occurred in combination with periodic floods and sediment inputs from a tributary.

Is it possible to manage a river as highly regulated as the Colorado in ways that protect and improve environmental conditions for the native biota? Only time will tell, but an important first step is recognizing that key processes and conditions must be allowed to fluctuate within a range of natural variability.

Challenges Ahead

The problems confronting freshwater ecosystems will be intractable if they continue to be approached piecemeal. Several government programs, such as the EPA Clean Lakes Program, the Wetlands Restoration Act, and even the Endangered Species Act, mandate actions to prevent specific aspects of ecosystem degradation. But these programs are narrow in focus, effectively addressing symptoms rather than root causes of aquatic ecosystem decline. Control of pollution is necessary, for instance, but insufficient for maintaining a native species community if adequate water flows are not available at the right time, if the channel has been severely degraded, or if invasive species have been allowed to take hold. The needs of aquatic ecosystems and the needs of society for water supplies must be addressed collectively if freshwater ecological integrity is to be maintained or restored. Politically, this requires that broad coalitions of water users must work together towards a mutually acceptable future.

The best time to develop such coalitions is before water is allocated and before ecological crises occur. In many parts of the world, this opportunity was missed long ago. The potential for full or partial restoration remains, however. An ambitious example is taking place in south Florida, where water control structures are being physically removed and nutrient inputs curtailed in an attempt to encourage a more natural system. Other restoration projects around the nation also show promise.

The ecological consequences that arise when freshwater ecosystems are deprived of adequate water, proper timing of flows, and suitable water quality often become apparent to people only after the degradation begins to interfere with societal uses of freshwater. Nuisance algal blooms and loss of commercial or sport fisheries are examples of failures in ecosystem processes that were often years in the making. Some ecosystems naturally experience wide swings in environmental and ecological conditions from one year to the next that can mask gradual changes in physical and chemical factors. Most systems are inherently resilient to a particular pattern of disturbance, and their plant and animal communities will persist as long as conditions fluctuate within a certain range. Once a threshold is reached, however, these ecosystems may change rapidly to a new stable state that is very difficult to reverse. The collapse of a fishery and permanent cultural eutrophication from nutrient inputs are two examples of conditions that, once reached, make it difficult to restore the integrity of a freshwater system. Detecting such trends before problems become critical requires both monitoring the biological and physical conditions in freshwater ecosystems and understanding the natural ecological dynamics of these systems.

Study of the Great Lakes Ecosystems

The Great Lakes – Superior, Michigan, Huron, Erie, and Ontario – hold 20 trillion cubic meters of freshwater, approximately 18 percent of the planet's freshwater supply. The overall basin is home to 35 million people, including 10 percent of the U.S. population and 25 percent of the Canadian population. Nearly 25 percent of agricultural production in Canada, and 7 percent of the agricultural production in the United States occurs in the basin. In addition, the Great Lakes provide drinking water for 40 million people and supply 210 million cubic meters of water per day for municipal, agricultural, and industrial use.

Poor water quality caused by excessive inputs of phosphorus and nitrogen is one of many serious problems affecting the Great Lakes. Some basins of the lakes also contain exceedingly high concentrations of toxic chemicals; habitat destruction has been significant and is increasing; native fisheries have been greatly altered or intentionally replaced; invasive species have altered native food webs and water quality and also damaged human infrastructure; and climate change is expected to alter lake levels. Although freshwater environments the world over share many of the same problems, their significance is heightened by the sheer size of the Great Lakes and the quantity and quality of their waters.

Water Quality. Water quality in the Great Lakes has improved dramatically from the eutrophic conditions that prevailed prior to the 1980s. This has been achieved through greater regulation of point-source pollution. However, water quality has not been restored to "natural condition." Years of phosphorus enrichment in Lake Michigan, for example, increased the growth of diatoms and depleted lake silica concentrations (silica is a necessary nutrient for diatoms and sinks to the lake bottom when diatoms die). Without enough silica, natural algal assemblages and the zooplankton that feed upon them have been severely altered. Today, cultural eutrophication may actually be masked by the filtering activity of zebra mussels, which increases water clarity by shifting nutrients from the water column to the lake sediments. Nonpoint source pollutants, including fertilizers, pesticides, sediment, and bacteria, still significantly impair Great Lakes water quality.

Invasive Species. Non-native species have modified habitats, reduced native biodiversity, and altered food webs. An estimated 162 exotic species now reside in the Great Lakes, including introduced sport fish. Although the zebra mussel and sea lamprey have received the most attention, many other less apparent species profoundly affect the ecosystem, including quagga mussels, predatory zooplankton such as *Cercopagis pengoi* and *Bythotrephes cederstroemi*, the benthic amphipod *Echinogammarus ischnus*, and the round and tubenose gobies. In addition to their ecological impacts, lamprey cost $10 million in control efforts each year, and zebra mussel control has totaled some $4 billion as of 2001.

Toxic Chemicals. The sediments in the Great Lakes store organic and inorganic contaminants coming from industrial, urban, and agricultural runoff as well as atmospheric deposition (including mercury and PCBs). Contaminants from sediments accumulate in aquatic species, affecting fish and wildfowl health and even the health of humans who eat contaminated fish. Contaminants also affect shipping, a major industry on the Great Lakes, because of potential restrictions on dredging of channels and harbors (which can release contaminants into the water column) and on disposal of dredged sediments.

Habitat Destruction. Land use changes have resulted in habitat loss throughout the Great Lakes basin. Urban sprawl continues to replace natural areas, farmland, and open space. The quality and quantity of coastal wetlands are declining; and the extent of hardened shorelines (that is, reinforced by sheet piling or rip rap) appears to be increasing, thus isolating wetlands from lakes, destroying habitat, and altering natural sediment movements.

Climate Change. Implications of future climate change in the Great Lakes region are profound. Some climate change models suggest conditions that will lead to lower lake levels, creating problems for the shipping industry as well as changes in water supply and environmental conditions in the lakes. Current climate models also suggest more extreme swings in climate, and unusually wet years may lead to periodic flooding. It is important to note that the 35 million people in the Great Lakes Basin are unprepared for large changes in lake level in either direction.

As an example of freshwater integrity, the Great Lakes fail on most accounts: shoreline hardening affects connectivity of the lakes with their wetlands; the current chemical and nutrient conditions represent a permanent change from natural conditions; and the plant and animal assemblages have been highly modified by human intervention. Constant effort and expense are now required to maintain water quality at acceptable levels, remove the legacies of past toxic inputs, control harmful non-native species, and restock valued recreational fisheries with exotic game fish that do not naturally reproduce in the lakes. Perhaps the Great Lakes can never be "restored" to the point where they are functionally self-sustaining, and therein lies a hard lesson. Many goods and services valued by society are no longer available (such as fisheries uncontaminated by toxins), and others are possible only through continuing expenditures of millions of dollars in remediation.

Balancing Human use and Needs of Freshwater Ecosystems

The sustainability of aquatic ecosystems can best be ensured by maintaining naturally variable flows, adequate sediment and organic matter inputs, natural fluctuations in heat and light, clean water, and a naturally diverse plant and animal community. Failure to provide for these essential requirements results in loss of species and ecosystem services in wetlands, rivers, and lakes. Aquatic ecosystems can be protected or restored by recognizing the following:

1. Aquatic ecosystems are not simply isolated bodies or conduits but are tightly connected to terrestrial environments. Further, aquatic ecosystems are connected to each other and provide essential migration routes for species.
2. Dynamic patterns of flow that are maintained within the historical range of variation will promote the integrity and sustainability of freshwater systems.
3. Aquatic ecosystems additionally require that sediment loads, heat and light conditions, chemical and nutrient inputs, and plant and animal

populations fluctuate within natural ranges, neither experiencing excessive swings beyond their natural ranges nor being held at constant, and therefore unnatural, levels.

Stating these requirements for maintaining aquatic ecosystem integrity, of course, is not the same as implementing them in the context of today's complicated society. U.S. water policy currently supports increased exploitation of water supplies in order to meet human demands. Policies for maintenance of water quality and flow are primarily based on human health needs. The age of ever-increasing exploitation is over, however. We must begin to redefine water use based on the recognition that supplies are finite and that healthy freshwater ecosystems must be sustained or restored. For these reasons we offer the following recommendations for how water is viewed and managed:

1. *Incorporate freshwater ecosystem needs, particularly naturally variable flow patterns, into national and regional water management policies along with concerns about water quality and quantity.* Because most land and water use decisions are made locally, we recommend empowering local groups and communities to implement sustainable water policies. A large and growing number of watershed groups is already moving in this direction with the support and guidance of state and federal agencies. Flexibility, innovation, and incentives such as tax breaks, development permits, conservation easements, and pollution credits are effective tools for achieving freshwater ecosystem sustainability goals.
2. *Define water resources to include watersheds so that freshwaters are viewed within a landscape or systems context.* Many of the problems facing freshwater ecosystems come from outside the lakes, rivers, or wetlands themselves. Laws and agency regulations lag in their recognition of this fact. One place to initiate a change is through existing governmental permitting processes. The EPA's TMDL Program is an effort to address both point and nonpoint pollution from a watershed to a water body, although the program has not yet been fully implemented. It should also be refined to consider how flow variability influences the transport of pollutants.
3. Increase communication and education across disciplines. Interdisciplinary training and experience, particularly for engineers,

hydrologists, economists, and ecologists, can foster a new generation of water managers and users who think about freshwaters as systems with ecological purposes as well as water supply functions.

4. *Increase restoration efforts for wetlands, lakes, and rivers using ecological principles as guidelines.* While some restoration has occurred, a greater effort is required to restore the ecological integrity of the nation's water resources. The goal of restoration should be to reinstate natural variations in the fundamental environmental factors identified above. Yet many restoration projects, especially for wetlands, have focused only on replanting vegetation while ignoring underlying hydrologic, geomorphic, biological, and chemical processes. Highly visible yet ecologically incomplete restoration efforts such as these wetland revegetation projects may even foster complacency among the public. A recent Gallup Poll found that Americans are increasingly satisfied with the nation's environmental protection efforts, making them less likely to support the funding and political effort needed to enact genuine ecological restoration requirements. In any given freshwater system, the extent of restoration and protection that is eventually undertaken will be widely debated because active management is inherently a social process, although one ideally informed by science. Restoration efforts can encompass a spectrum of goals, from nearly full recovery of native species and environmental conditions to the management of dynamic, biologically diverse communities that do not necessarily resemble native ecosystems.

5. *Maintain and protect remaining minimally impaired freshwater ecosystems.* Aldo Leopold said: " If the biota, in the course of aeons, has built something we like but do not understand, then who but a fool would discard seemingly useless parts? To keep every cog and wheel is the first precaution of intelligent tinkering." Many restoration projects fail to reestablish ecosystem functioning once major processes have been disturbed. It is far wiser and cheaper to conserve what we have. Moreover, our remaining functionally intact freshwater ecosystems can provide a source of plant and animal colonists for restoration projects elsewhere.

6. *Bring the ecosystem concept home.* Achieving ecological sustainability requires that we come to recognize the interdependence of people and the environments of which a part. For freshwaters, this will require

broad recognition of the sources and uses of water for societal and ecological needs. It will also require taking a much longer view of water processes. Water delivery systems and even dams are developed with life spans and management guidelines of decades to, at most, a century. Freshwater ecosystems have evolved over aeons, and their sustainability must be considered from a long-term perspective. Governmental policies, mass media, and a market-driven economy all focus on much shorter-term benefits. Educational programs at the kindergarten through high school level, individual initiatives to become informed, and efforts by local watershed groups interested in protecting their natural resources can provide good first steps toward enduring stewardship.

Restoring Freshwater Ecosystems in South Florida

The south Florida ecosystem covers approximately 47,000 square kilometers and ranges from Orlando in the north to the Florida Keys at its southern extreme. It includes the Kissimmee River, Lake Okeechobee, The Everglades, and Florida Bay. The landscape is essentially flat; the elevation drop from Lake Okeechobee to Florida Bay, a distance of 160 kilometers, is less than 6 meters. South Florida has undergone enormous changes in population, land use, and hydrology over the past 100 years, resulting in profound changes to ecosystem structure and functioning. Starting in the early 1900s, efforts were made to drain the Everglades wetlands, which were viewed as wastelands and useless swamps. Hurricanes and floods prompted massive water management projects. There are now more than 2,500 kilometers of levees and canals, 150 gates and other water control structures, and 16 major pump stations. The flood control system has worked remarkably well, making the region less vulnerable to the extremes of flooding and drought by storing water for supply and moving it for flood control. These management projects were designed in the 1950s when it was anticipated the population in the region would reach 2 million by the year 2000. Today, however, the region is home to more than 6 million people. More significantly, the water projects were not designed with environmental protection or enhancement in mind.

Environmental problems unintentionally created by these water management projects include:

(1) Up to 6.4 billion liters per day of excess rainwater is channeled directly to the ocean to keep urban and agricultural lands from flooding, causing

salinity imbalances in estuaries and influencing plant and animal communities.

(2) Lake Okeechobee is treated as a reservoir for water supply or flood control instead of as a natural lake.

(3) Water supply and periodicity for the Everglades has been altered, greatly harming the biota.

(4) And Water quality has deteriorated throughout the region.

Accelerated eutrophication of Lake Okeechobee from phosphorus runoff associated with dairy and beef cattle operations, for example, has shifted the composition of the algal, invertebrate, and higher plant community and thus, the food web. Phosphorus enrichment of the northern Everglades from sugar cane farms has changed the structure and biomass of the periphyton community (organisms attached to submerged substrates) while increasing cattails at the expense of sawgrass. Increases or decreases in the discharge of freshwater to estuaries have influenced the natural salinity patterns of these systems, affecting the abundance of seagrass, oyster, and fish communities. Channelization of the Kissimmee River caused the loss of 11,000 hectares of floodplain habitat.

Approximately half of the historic Everglades has been converted to agricultural or urban use. Populations of wading birds have been reduced 85 to 90 percent. Sixty-eight species of plants and animals in south Florida are threatened or endangered, and invasive species such as melaleuca, Brazilian pepper, Australian pine, torpedo grass, Old World climbing fern, and Asian swamp eel are threatening native habitats and species.

Although it is not possible to restore this region to its pristine condition, efforts are underway to redesign the south Florida aquatic environment to make it more compatible with the way the system formerly functioned. Congress has funded efforts to develop a Comprehensive Everglades Restoration Plan, an ambitious and innovative partnership that aims to enhance the region's ecological and economic values as well as the well-being of its human population. The objectives are to increase the amount of water available by storing it instead of sending it out to sea, ensure adequate water quality, and reconnect the parts of this ecosystem that have been disconnected and fractured. A multi-faceted approach has been proposed that may take 25 years or more to implement.

The ecological goals of the plan are to increase the extent of natural areas, improve habitat and functional quality, and improve native species richness and biodiversity. Success will be evaluated with quantitative criteria. For example, a goal for Lake Okeechobee is to reduce total phosphorus in the water column from a current concentration of 110 to 40 µg/L. Rigorous programs of scientific research will continue throughout project implementation in order to address major uncertainties. The information generated, combined with results from monitoring networks, will be used in adaptive management of the restoration plan.

References

Council on Environmental Quality (1995). *Environmental Quality*. 1994-1995 Report. Office of the White House, Washington D.C.

Daily, G.C., ed. (1997). *Nature's Services: Societal Dependence on Natural Ecosystems.* Island Press, Washington, D.C.

Environmental Protection Agency (1998). *National Water Quality Inventory: 1996 Report to Congress*. U.S. EPA EPA841- R-97-008, Washington, D.C.

Echeverria, J.D., P. Barrow, and R. Roos-Collins. (1989. *Rivers at Risk: The concerned citizen's guide to hydropower.* Island Press, Washington, D.C.

Naiman, R.J., and M.G. Turner (2000). A future perspective on North America's freshwater ecosystems. *Ecological Applications* 10:958-970.

National Research Council. (1992). *Restoration of aquatic ecosystems: science, technology, and public policy*. National Academy Press, Washington, D.C.

5

Comprehensive Assessment of Urban Freshwater Systems

Broadly defined, a sustainable city embodies all the good qualities a city can have, and perhaps a few more besides. A healthy city is much the same. Yet the goals of sustainability and health, more narrowly defined, are by no means identical. And however valiantly one tries to extend the concept of the sustainable city to cover contemporary deprivations, it implies a special concern not to compromise the future. Likewise, however valiantly one tries to extend the concept of a healthy city, it implies an emphasis on traditional environmental health concerns.

In looking at urban water issues, it adopts a narrow usage. Perfectly sustainable water systems would, according to this narrow definition, be indefinitely maintainable, and would not undermine the ecological systems and natural resources upon which the city depends. Perfectly healthy water systems would minimise water related diseases, and meet all of the residents' basic water needs. Whether a sustainable water system is healthy, or a healthy one sustainable, is left open. This allows two of the main challenges for 20th century urban water systems to be kept conceptually distinct, and avoids conflating interrelation with identification. While it is important to try to make the pursuit of sustainability and health complementary, it is equally important not simply to assume conflicts away.

Water is universally recognised as a critical resource, but the sustainability of water supplies is often overlooked in planning for a city's future. This is not a peculiarly modern problem. Writing in the 14th century on "Requirements for the planning of towns and the consequences of neglecting those requirements," Ibn Khaldûn placed water for human consumption first in the list of resources whose negligence had made a number of Arab cities of the past "very ready to fall into ruins, inasmuch as they did not fulfil all the natural requirements of towns". His concern is echoed in modern texts, describing cities straining the limits of their water supplies. Modern technologies, with their much greater capacity to draw on distant supplies, shift but do not eliminate urban water supply constraints.

In the last two centuries, urban growth and industrialisation have also added to the range of pollution burdens which can threaten the ecological sustainability and health of cities. Many coastal zones, rivers and downstream inhabitants have been gravely damaged by urban polluters who use water to carry away hazardous pathogens or chemicals. This strategy of displacing urban environmental burdens onto the water systems has also long been questioned, perhaps most eloquently by Coleridge, who penned the following lines at the end of a poem on Cologne in 1828:

> The river Rhine, it is well known, Doth wash your city of Cologne; But tell me, Nymphs, what power Divine Shall henceforth wash the river Rhine?

The Rhine today is less central to the "washing" of cities like Cologne (though some would say the burden has yet again been displaced rather than removed), and the quality of the water in the Rhine has improved as a result. However, more populous, more industOrial, less wealthy cities are polluting their waterways on a far larger scale, and Coleridge's question still resonates.

Nineteenth century commentators also criticised the manner in which new urban technologies, and water borne sewerage systems in particular, interrupted the "circulus" of nature. Again, it is even possible to find a literary reference, this time in Victor Hugo's *Les Misérables*, where he laments the "blindness of a bad political economy" which allowed faeces to be carried off with water rather than used in the fields. Also an advocate of assigning monetary values to environmental problems, he claimed that: "Each hiccup of our cloaca costs us a thousand francs. From this two results: the land impoverished and the water contaminated. Hunger rising from the furrow and disease rising from the river".

If seemingly modern issues of urban water, nutrient cycles and sustainability have actually been with us for centuries, seemingly old fashioned issues of water, poverty and infectious diseases may unfortunately be with us for centuries to come. Reports based on government statistics indicate that in the early 1990s about one fifth of urban dwellers in the South were without adequate access to safe water and about one third were without adequate sanitation, and what governments define as adequate may still be far from healthy. There are roughly three million deaths every year from diarrheal diseases, most of which could probably be averted by the better use of water. While water related health problems are likely to remain far more severe in the South, some are beginning to re-emerge in wealthier countries. The 1993 diarrhoea outbreak in Milwaukee in the American Midwest, where more than 4,000 were hospitalised after being infected with a parasite in the municipal water supplies, was a challenge to Northern complacency.

The current burden of water-related diseases in urban areas is not, by and large, the outcome of the city-wide water supply and pollution problems that threaten sustainability. Low income urban neighbourhoods and households are more likely to lack water because they cannot access the city's water supplies than because those supplies are limited. A comparatively healthy overall water balance can be accompanied by extremely unhealthy conditions in disadvantaged neighbourhoods. Indoor piping and low water prices may be the rule in one neighbourhood, while residents of a squatter settlement nearby must choose between a heavily polluted stream and extremely expensive water from vendors. Water may be clean in the pipes, but heavily contaminated by the time it has been carried home, stored, and ladled into a guest's drinking vessel. For many of the more disadvantaged urban dwellers, water scarcity and life-threatening water pollution are not future prospects given unsustainable practices, but current realities. Only concerted actions, based on better science, more respect for local knowledge, and politics that create a continuous pressure for improvement, are likely to make significant headway.

Alternatively, a city's water future may be grim even when cheap and reasonably clean piped water remains widely available. Indeed, many of the measures employed to address urban environmental health problems, such as providing low-price piped water supplies and water borne sewerage systems, can add considerably to the water-environment burden of a city. Thus the water use patterns of middle income mega-cities are particularly unsustainable, despite the fact that the average prevalence of water-related diseases is likely to be lower than in smaller but poorer urban centres.

As such, the health and sustainability aspects of urban water systems are both important, but worth keeping conceptually distinct. In wealthy cities where piped water and water borne sanitation is universally provided, sustainability can be a very useful organising principle for water system improvement. Improvements should not be a threat to health, of course, but it is in the domain of sustainability, more narrowly defined, that most progress needs to be made. Alternatively, in the more disadvantaged neighbourhoods, where water scarcity and quality problems are a serious and immediate threat, the water agenda ought to be developed around a concern for health and welfare. Improvements should be sustainable, of course, but sustainability is not appropriate as an overarching goal, however progressive this may seem.

Perhaps equally surprisingly, when capital intensive solutions are not an option, as is the case in most areas where sanitary conditions remain a serious problem, science still provides very limited guidance for water and sanitary improvement. The tendency to focus on a few simple threats, such as contaminated drinking water, has helped obscure a broader ignorance. There are a number of both scientific and institutional issues to resolve before the "brown" agenda can justifiably be called an "easy" agenda. More efficient finance and infrastructure alone will not be sufficient.

The urban growth of this century, with the appearance of mega-cities as well as an unprecedented number of, much less noticed, smaller cities, is central to future urban water development. Linear material flows which underpin urban life allow urban dwellers to abuse natural resources and displace the burdens of both pollution and resource scarcity to faraway natural and human habitats. These processes too are poorly understood, and pose a range of seemingly intractable institutional problems. While some particular threats are well understood, it will not be enough to tackle just these. Just as poverty does not explain water-related ill health in low income cities, so affluence does not explain the water-related threats to sustainability that urban economic development can bring. On the other hand, achieving sustainability while maintaining urban affluence is inherently difficult.

In many contexts there are trade-offs between health and sustainability. Crude policies for improving health through water supplies, such as price controls and the extension of capital intensive water borne sewerage systems, tend to bear a high sustainability burden. Crude policies for improving sustainability, such as instituting high water prices and letting the market "decide" how the water should be allocated, tend to bear a high health burden. Generally, however, better informed, more efficient, and less politically compromised efforts to improve health or sustainability create fewer trade-offs, and can even make the pursuit of these goals complementary.

Water And Urban Health

Sanitary Movement, Water Science and Water Engineering

The sanitary movement was in many respects the environmental movement of the 19th century: it brought attention to hazardous side effects of urban development; it engaged public figures as well as scientists in debating unresolved scientific, political and moral issues; and it changed the course of urban development. Despite a lack of scientific consensus, and considerable pressure to do nothing to interfere with economic growth, a great deal was accomplished, especially in affluent cities. To some degree, problems were displaced rather than resolved. But a health transition was accomplished along the way, and the level of collective action achieved bodes well for the contemporary environmental agenda. On the other hand, this history also raises a difficult question for the environmental movement. Have the 'old fashioned' problems of water, local sanitation and health lost their prominent place on the environmental agenda because they are less important globally, because our environmental horizons have expanded, or because these 'old fashioned' environmental problems are no longer critical in the affluent parts of the world?

Most 19th century European cities were very unhealthy places to live. Mortality rates were far higher in the rapidly growing urban centres than in the surrounding countryside. In many cities mortality rates exceeded birth rates, and only rapid rural-urban migration allowed them to expand. The poor quality of the drinking water and the lack of wash-water must take at least part of the blame.

Sanitary Movement and Broader Environmental Concerns

Looking back, it is easy to view the sanitary improvements of the 19th and early 20th century as feats of science and engineering, now perhaps rendered less glamorous by an awareness of their broader environmental inadequacies. At the time, one might argue, scientists were just discovering that the routes of disease transmission involved human waste and microbially contaminated water. When engineers devised means for cutting off many of these routes with piped water, water quality control, and sewerage systems, they had to rely on existing science. It was hardly surprising that they failed to consider the consequences of introducing water borne sewerage on, for example, nutrient cycles. Such sophistication had to await late 20th century environmentalism.

This interpretation is seriously misleading. Partly because human waste was widely used as a fertiliser, a loss was quickly perceived when it no longer got to the fields (the intent of many of the water borne sewerage systems was originally to deposit waste on farmers' fields). The debates concerning sanitary improvement in the 19th century were as wide ranging as the environmental debates of today. And in any case, many of the major public health improvements were undertaken before the bacterial theory of disease was established, and scientific opinion on the relation between water and health was deeply divided. The theories that lost out in the course of scientific advance were often as supportive of sanitary improvement as those which won. For example, the once reputable scientific view that ill health was brought on by miasmas, a sort of air pollution that could disturb the balance of humours in one's body, was often used to justify efforts to improve sanitary conditions. Indeed, some historians have argued that miasma theory, with its holistic tendencies, provided more support for general cleanliness and the removal of filth, than the bacterial theories which tended to place the emphasis on particular pathogens.

Sanitary Movement and Public Debate

Furthermore, sanitary improvement in the 19th century was not the preserve of experts it was to become in the early 20th century. Just as in contemporary environmental debates, scientists were brought in by all sides, to justify both environmental intervention and *laissez faire*. And at the height of the public health movement, debates about water and health engaged not just scientists and engineers, but a whole panoply of public figures. Indeed, there was considerable popular interest in sanitary improvement. The General German Exhibition of Hygiene and Life Saving held in 1883 attracted 900,000 people to Berlin over the course of five months. It was only later, when a conventional approach to water and sanitation had been established, that public interest in such events waned, and experts took over most of the key positions. As Hamlin notes, we are again entering a period of uncertainty about water, and must again learn to debate water issues in public. In this context, it is important to take the right lessons from the 19th century experience.

Ironically, one of the dangers of treating the sanitary revolution as a scientific and technical achievement is that, even as it exaggerates the importance of science in the past, it ascribes an unduly limited role to science in the future. This applies especially to the notion that an understanding of microbial disease transmission brought about the sanitary revolution, and more generally the epidemiological transition away from infectious diseases in the North. Such accounts not only overstate the significance of such scientific breakthroughs as John Snow's demonstration of the link between cholera and drinking water, but also mask the profound ignorance which remains regarding the transmission and evolution of water related diseases. Certainly a great deal has been learned, but not so as to provide any easy answers.

Sanitary Improvement as a Technical Fix

Once economic growth and a set of public health measures, including piped water and sewerage, were found to reduce the incidence of most water related diseases to comparatively insignificant levels, the motivation to understand them better fell drastically. Debates about which specific measures underpinned the epidemiological transition seemed academic when the whole package was clearly desirable. Moreover, as practitioners find in so many sectors, funding for environmental service utilities was easier to justify when uncertainties were ignored. Even some people working in the water sector came to believe that getting people clean drinking water was the key to reducing water borne diseases. More generally, piped water and articulated sewerage systems came to be viewed as the technical fix for the urban health disadvantage. For the greater part of this century, the accepted wisdom has been that the urban health-related goal in the water and sanitation field should be to provide every dwelling with piped water and a water closet, or the closest possible equivalent. Indeed this was very much the premise of the Water and Sanitation Decade, which had as its global target "to provide all people with water of safe quality and adequate quantity and basic sanitary facilities by 1990" in. However, for many urban dwellers in the South, such technical fixes remain as elusive as ever. Public investment in capital intensive infrastructure is becoming less rather than more prevalent, and the private sector is unlikely to fill such a large and costly gap. With less costly systems, there is still enormous uncertainty surrounding the long-standing environmental health issues relating to water and sanitation. Once one leaves the areas where water comes out of indoor taps and faeces are flushed away with the press of a button, there is still the need for a holistic approach to environmental health issues involving water. There are numerous local externalities, which relate local water supplies with sanitation, insects, waste disposal and a host of community-level environmental problems. There is clearly a need for local collective action, articulated and supported both nationally and internationally. In short, there is a need for a new, less patrimonial, more sophisticated environmental health movement.

Urban Water and Modern Environmentalism

Northern environmentalists have recently managed to broaden international policy debates to include water and sustainability, introducing a number of complex and poorly understood challenges in the process. The Global Freshwater Assessment reflects this success. It is likely that today's theorising about sustainability will one day seem as quaint as the miasma theory of disease. Indeed, the notion that filth and impurity give off a miasma that can disturb the balance of humours in the human body is not so very different from the concern that cities disturb the water system and hence the balances of nature. Miasma theory helped underwrite an active holistic approach to sanitary improvement at a time when reductionist science was in danger of fostering inaction. Sustainability debates are hopefully providing much the same for broader environmental concerns. Unfortunately, however, they are in danger of further marginalising the environmental health concerns that spurred the urban based sanitary movement of the 19th century, despite the large numbers of people for whom these concerns are still central.

Water and Infectious Diseases

Overall, the situation regarding environment and health in disadvantaged neighbourhoods is not unlike the typical situation encountered in regard to the broader sustainability problems globally: it is far easier to make a long list of preventative steps which taken together would undoubtedly reduce the environmental health burdens considerably, than it is to demonstrate the importance of particular hazards and prioritise particular measures. Water is critical to the transmission of many diseases. In arguing for the importance of water, it is common to oversimplify its role, and overemphasise the significance of contaminated drinking water. Actually, the role of water in washing pathogens away from the path of potentially infected people is at least as important as its role in bringing pathogens to people. As a result, in areas where faecal-oral diseases are endemic, how much water people get, and how they use it, is probably more important than its quality.

There is still enormous uncertainty, however, concerning how faecal-oral diseases are most often transmitted, and which interventions are likely to make the most difference. Moreover, the accumulation, flow and quality of open freshwater is critical to the spread of malaria, dengue fever and a variety of other vector borne diseases. Here too, better quality water is not a sufficient remedy, and if pursued unthinkingly can even facilitate disease transmission. In combating water related diseases, it is possible to identify a few measures, such as hand-washing and applying oral rehydration therapy that are sufficiently general and important to advocate widely, but for the most part the simplest solutions come at a very high cost. Overall, while far more could be done on the basis of existing knowledge, a better understanding of these processes could still make an enormous difference.

Misleading Simplification

The message is simple, and seems to confirm what every world traveller experiences - bad water makes people ill, and there is a lot of bad water about. And yet water contamination is only one, perhaps quite small, aspect of water related disease transmission. The statistic of 80% probably derives from the enormous number of episodes of diarrhoea which are estimated to occur every year: about 4 billion in 1995. Indeed, diarrhoea still competes with acute respiratory infection for the position of principal childhood killer. Thus the more serious concern is that most of these statements falsely suggest that these diseases are the result of water contamination. In fact, water relates to disease in a wide variety of ways, with water contamination only one of many aspects.

Water Related Transmission Routes for Infectious Diseases

It presents a classification of water related transmission routes. With the *water-borne* route, water brings the pathogens that people ingest to become infected - in such cases contaminated water really is the culprit. With the *water-washed* route, people become infected because water failed to carry the pathogens away - it is the absence of washing that is used to define the route. *Water-based* transmission refers to infections whose pathogens spend part of their life cycle in aquatic animals. Transmission via *water-related insect vectors* refers to diseases spread by insects that breed in or bite near water.

Table 1. Environmental classification of water-related infections (I), and the preventive strategies appropriate to each

Transmission route	***Preventive Strategy***
Water-borne	Improve quality of drinking water
	Prevent casual use of unprotected sources
Water-washed	Increase water quantity used
(or water-scarce)	Improve accessibility and reliability of domestic water supply
	Improve hygiene
Water-based	Reduce need for contact with infected water
	Control snail population
	Reduce contamination of surface waters
Water-related insect	Improve surface water management
vector	Destroy breeding sites of insects
	Reduce need to visit breeding sites
	Use mosquito netting

Classifying Water Related Diseases

The relation between transmission routes and types of disease is far from perfect, although similar classifications have long been applied to water-related infections. What is presented in a variation of the classification developed by David Bradley in the early 1970s, upon which most current classifications are based. The original terminology distinguished water-borne, water-washed, water-based and water-related infections, often leading people to conflate the possibility that a disease may be borne by water with the much stronger claim that it is usually or always borne by water. In fact, the distinction between water-borne infections and water-washed infections has always been problematic, as Bradley has made very clear:

> "All infections that can be spread from one person to another by way of water supplies may also be more directly transmitted from faeces to mouth or by way of dirty food. When this is the case, the infections may be reduced by the provision of more abundant or more accessible water of unimproved quality".

The tendency to ignore the considerable uncertainty that still surrounds these infections and how they spread often creates the impression that expert opinion is vacillating. The following statement in the World Health Report 1996 illustrates this seeming instability:

> "It was long thought that contaminated water supplies were the main source of pathogens causing diarrhoea but it has now been shown that food has been responsible for up to 70% of diarrheal episodes"

Unfortunately, such statements suggest that the remaining uncertainties are minimal, though with terms like "up to," any real commitment to accuracy is avoided. In fact, the relative importance of the different routes *faecal-oral* diseases can take remains in large part a mystery (and need not always involve a *faecal-oral* route). Nor is this simply a matter of not knowing in sufficient detail what conditions are like in the more disadvantaged neighbourhoods. Conditions are sufficiently bad in most under-serviced neighbourhoods that any number of possible transmission rates could explain a high level of faecal-oral disease. Indeed, experts are often left wondering why the prevalence of faecal-oral diseases is not higher in many neighbourhoods.

Flies play a potentially important but largely unknown role in the transmission of faecal-oral diseases. Faecal pathogens are known to flow via groundwater from pit latrines to nearby wells, but there is enormous uncertainty in how far is safe. Numerous behavioural patterns can affect the likelihood that people, and especially children, will directly encounter faecal material and ingest it, but these are difficult to identify let alone prioritise.

The paths of transmission which particular *faecal-oral* pathogens will actually favour depend on a variety of features, such as the infectious dose, persistence, and multiplicability in food, and some paths may to lead to people likely to be immune. The measures which can be taken to prevent transmission also depend upon where the transmission takes place, and it can be important to distinguish between transmission within the domestic domain as opposed to transmission in the public domain. Moreover, people may engage in defensive behaviour when the hazards are perceptible, which itself changes the effects of certain hazards, and needs to be taken into account when designing policy responses.

> Although no source is given, it seems likely that the upper limit of 70% is from a WHO commissioned paper which estimated that "On the basis of the predicted impacts of controlling transmission associated with poor personal hygiene, and water supply and sanitation, food as a vehicle may

> contribute to the transmission of between 15 and 70% of all diarrhoea episodes. This range is so wide as to be of little value, but the available data do not permit more precise estimates".

Water Quality Versus Water Quantity

Looking specifically at the alternatives of improving water quality or improving people's access to sufficient quantities of water as a means to reducing the transmission of diarrheal diseases, there is a good case to be made for claiming that too much attention is given to water quality. It is difficult to find clear epidemiological evidence distinguishing between the effects of improving water quality and providing better access to water, but while a number of studies have demonstrated the importance of providing more water to poor households, the evidence with respect to water quality is more ambiguous.

It has been suggested that one of the reasons many studies fail to find a significant association between improvements in water quality and diarrhoea prevalence, is that most contamination occurs in the home while tests for faecal contamination are typically made at the well or tap. Indeed, a number of studies have documented higher concentrations of faecal coliforms in household water containers than at their sources.

However, it is far from certain that contamination that occurs in and around the home is comparable to contamination from outside. A recent and influential study contends that "In-house contamination does not pose a serious risk of diarrhoea because family members would likely develop some level of immunity to pathogens commonly encountered in the household environment. Even when there is no such immunity, transmission of these pathogens via stored water may be inefficient relative to other household transmission routes, such as person-to-person contact or food contamination. A contaminated source poses much more of a risk since it may introduce new pathogens into the household". However, the empirical evidence in support of these claims is very limited, and the argument itself depends heavily on some behavioural assumptions one would expect to be culture dependent. If, for example, children play freely in the neighbourhood and drink in the homes of neighbours, then a sharp distinction between in-house and communal water contamination is not convincing.

Most of the complications mentioned up to this point suggest that water quality receives too much attention relative to the role of water in washing away pathogens. Another generally ignored complication, the evolutionary implications of different water systems, suggests a rather different conclusion. In placing obstacles on some of the routes of disease transmission, not only is the rate of transmission altered, but so are the evolutionary pressures brought to bear on the pathogen. It has been argued that interrupting water-borne routes will put in motion evolutionary pressures towards less virulent strains of, for example, cholera. To put it crudely, pathogens that rely more heavily on personal contact lose out if their hosts are immobilised, while those that are *water-borne* do not. Thus, so the argument goes, eliminating the water-borne route creates evolutionary advantages for strains that are not debilitating. Ewald derives estimates that suggest that taking account of evolutionary pressures would radically alter the benefits ascribed to water purification in Bangladesh, turning it from an uneconomic alternative to a measure costing one tenth as much per life saved as vaccination. Thus, rather than immunising the potential hosts, one would leave larger scope for less virulent pathogens, to the disadvantage of the more dangerous strains.

Another role water has to play in faecal-oral diseases lies in their treatment. Dehydration is usually the proximate cause of death in fatal cases of diarrhoea, and can often be treated with water and oral rehydration salts. Initially developed in Bangladesh, this simple measure has been heavily promoted by UNICEF, and it has been estimated that around half of diarrhoea cases are now treated with oral rehydration therapy.

Water and Insect Vectors

Other than faecal-oral diseases, the most important category of water related disease is the last: water related insect vector. Malaria, filariasis, yellow fever, and dengue fever are all classified as *water related diseases* because their vectors breed in water. Malaria alone kills an estimated 1-2 million people annually, and there are up to half a billion new cases every year, second only to diarrheal and respiratory infections. The role of water contamination in the transmission of these insect borne diseases is very ambiguous, however. Indeed, one of the reasons malaria is typically more prevalent in rural than in urban areas is that most malarial mosquitoes, like humans, prefer clean water. Urban water pollution can actually help protect residents from malaria, though some malarial species have found urban water niches. The mosquitoes which carry dengue haemorrhagic fever have generally been more successful in finding urban water niches, and often breed in household water containers. Such containers tend to be more common where water supplies are intermittent, but are only indirectly linked to water contamination.

Despite the complexity of these various water related diseases, in affluent cities they are only rarely a problem. While experts may debate exactly how and why this epidemiological transition takes place, inadequate sanitation in the broad, 19th century sense is closely associated with poverty. Poverty alleviation would seem to be one of the most obvious means to achieve improvements in environmental health. Indeed, one possible justification for leaving the issues of water, sanitation and health off the environmental agenda is that the solution is poverty alleviation and infrastructure provision, not environmental management. For a variety of reasons, however, such justification is weak at best.

Water, Poverty, Illness and the Need for Collective Action Water, Sanitation and the New Environmental Agenda

Growing or new environmental problems command more attention than declining or endemic problems, for obvious reasons. One obvious reason is that declining and endemic problems seem to be 'under control,' while growing problems give the impression of 'getting out of hand.' A second obvious reason is that the articulate, powerful and wealthy have typically already managed to avoid the declining problems, but are often potentially at risk from the new and growing problems. Both of these reasons help to explain the fact that the longstanding problems of water and health sit rather uneasily on the contemporary environmental agenda, which is still dominated by the concerns of the more affluent North.

Water and Sanitation Coverage and Expansion

While the long-standing problems of water and health affect a declining share of the urban population, they remain extremely severe, especially in lower income cities and neighbourhoods. It is of little consolation to those without to learn that others now have hygienic toilets and good water supplies.

However, the optimistic international statistics often seem to be at odds with local studies, and there are reasons to believe that such coverage figures exaggerate the progress to date. The international statistics are based on government estimates, with very little guidance given on what constitutes adequate water and sanitation. Such statistics, and especially their progression over time, are taken to reflect on government performance. Thus there is continuous, and often increasing, pressure to interpret 'adequate' liberally.

Increasing Emphasis on Financial and Environmental Sustainability

In the wake of the Water and Sanitation Decade, the dominant approach and the goals of water and sanitary improvement changed. In the influential Dublin Statement of the International Conference on Water and the Environment of January, 1992, the target year for universal coverage for all was shifted back to 2015

The Resolution 35/18 of the General Assembly of the United Nations actually only "*Proclaims* the period 1981 - 1990 as the International Drinking Water Supply and Sanitation Decade, during which Member States will assume a commitment to bring about a substantial improvement in the standards and levels of services in drinking water supply and sanitation by the year 1990", and cost recovery was promoted more vigorously than service coverage. This reflects a more general shift in emphasis and approach. Both physically and organisationally, more emphasis is now put on sustainability. Physically, more attention is being paid to ensuring that ecological damages are taken into account. Organisationally, more attention is paid to ensuring that the water provision is financially viable. Improvements are sorely needed in both of these areas. On the other hand, it is important that more attention to ecology and finance does not mean less attention to health and environmental justice.

The current emphasis on financial sustainability in water provision has developed in tandem with a more market-oriented approach to service provision. The promotion of simple organisational approaches to water and sanitation provision has long proved as attractive as the promotion of simple accounts of how faecal-oral diseases spread. The notion that governments could simply plan their way to improved water and sanitation was clearly misguided. To some degree, recent advocacy of cost-recovery through user fees and private sector involvement is yet another enticingly simple prescription, that aims to correct some real flaws in existing practice, but will become pernicious if it is widely accepted and adopted as a new orthodoxy.

Shifting Responsibilities Between Public and Private Sectors

Last century, the notion that by freeing up markets one could provide the solution to water and sanitary problems would have seemed almost perverse. Markets were far more pervasive in the urban areas than in the more tradition bound countryside, and urban filth and high mortality rates were associated with the market-oriented cities. Experience with unregulated private action helped to justify the public health movement, and eventually the view that the government had to provide urban water and sanitary services. But just as market failures once helped to justify public control, more recent public failures have helped justify a shift back towards private sector solutions. Unfortunately, shifts back and forth along the public-private axis do not represent progress.

In both the public and private sectors there are serious problems creating and maintaining the right incentives with respect to water and sanitation. Efforts to promote one or the other sector often involve exaggerating the problems of the opposing sector and ignoring or arguing the problems in the favoured sector away. Thus, private sector proponents tend to ignore the fact that, especially in low income areas, water and sanitation involve a host of externalities and public goods problems. Similarly, public sector proponents have long ignored the fact that subsidies justified on public health grounds are often systematically diverted to wealthy households and neighbourhoods where public health risks are minimal.

Re-examining the Allure of Centralised Water and Sanitation Systems

The physical complexity in the relation between water and health, as summarised in the previous section, involves a series of environmental interrelations. These interrelations make it inappropriate to treat water and sanitation as normal, partible economic goods, which ought to be bought and sold in the same manner as, for example, bicycles or rice. If water is allowed to accumulate on someone's land and become a breeding ground for mosquitoes, it is not just the water user that is affected. If water is used to wash faecal material into a street-side gutter, it is not just the water user that is affected. And if water is not used to maintain hygiene and a child falls sick, the child's playmates are placed at risk. Where such practices are or could become common, water prices cannot provide sufficient signals on how and when water should be used. And these practices are potentially widespread in most deprived neighbourhoods without piped water and sanitation systems. As well as being convenient for the careful user, one reason for the success of traditional piped water and water borne sewerage systems has been that, in the right circumstances, even the careless find it more trouble than it's worth to allow water to accumulate, flush sewage into a gutter, or generally engage in unhygienic practices that threaten public health. The piping itself severs many of the more hazardous environmental interconnections. The water leaves the piping only to enter the drains, often seconds after it emerged. At least in the technologies' centres of origin, all that is required of the user is a few norms which, once adopted, are easier to follow than to flout. To collect and store water in a bucket, and efficiently use small quantities of it, demand a far better understanding of the physical processes and the health risks involved than does running a modern household. Indeed, many environmental services such as piped water and sewerage connections not only displace pollution problems, but also shift both the intellectual and practical burdens of environmental management from the households to the government or utility company.

Another attraction of piped water and sewerage has been that the resulting system is amenable to relatively tight, centralised control. Engineers can design the system, and a water utility can manage it. Technologically and environmentally, using water to carry away human waste has numerous serious drawbacks. But, being well designed for centralised control and management, the piped systems are in other ways a civil engineer's and a civil servant's dream.

Practical Obstacles to the Public Provision of Water and Sanitation

In a great number of urban areas, however, universal coverage with piped water and sewers is just that: a dream. In very low income settlements, it is simply unaffordable, for either residents or their governments. Moreover, centralised control, and the desire to retain it, can easily become a liability when public finance is inadequate, good governance is problematic, or the government simply does not respond to the needs of the currently unserved. Such conditions are all common, particularly when both public and private finance is very scarce. Often, the result is low priced services that the utility can not afford to extend. As a result, the subsidies which do exist often go to the wealthy who, even without the subsidy and given the alternative of having a poor service at a low cost, would choose pay the full economic cost for adequate services. Those people who would need subsidies to be convinced to pay for healthy water and sanitation often end up having to pay exorbitant prices for inadequate services. In a somewhat perverse manner, the fact that utilities are forced to charge low prices often precludes them from expanding their coverage, and those not connected end up paying market prices for artificially scarce water and sanitary facilities.

Local Collective Action and Water and Sanitary Improvement

Given the problems getting either private enterprises and public utilities to provide suitable water and sanitary services to low income communities, it is tempting to look to NGOs and CBOs to provide the organisational solution. Indeed, many of the more sophisticated attempts to promote a market oriented approach to water and sanitation also emphasise the importance of local participation, which eventually must mean some form of local organisation. As described above, many water-related diseases involve spatially delimited public goods. Where such diseases are endemic, one's health depends very much on the water and sanitation practices of other member's of one's household and neighbourhood. The built facilities (e.g. wells, standpipes, latrines, condominial sewers) are often shared. If neither the public sector nor private enterprise can secure healthy local water supplies and sanitation, perhaps the responsibility should be given to the local communities and CBOs, working with the support of NGOs.

As long as the public sector is providing subsidised water and sanitation in wealthy areas, advocating unsubsidised participatory solutions in low income neighbourhoods is inequitable and probably inefficient. Indeed, however well the local communities managed to solve their water and sanitation problems, the economic signals would be perverse: subsidies where resource externalities predominate and economics indicates a need for taxes, and no subsidies where health externalities predominate and economics suggest subsidies are warranted. Moreover, there is a serious danger that attempts to roll back the state and support local initiatives, will succeed in the former and not in the latter, leaving already disadvantaged communities even worse off.

There are undoubtedly a number of good reasons to support local initiatives, and what could loosely be termed "community empowerment" for water and sanitary improvement. Indeed, the arguments for greater local participation and empowerment extend to other aspects of environmental management and urban management generally. However, just as the abstract notions of the perfect market and the perfect planner should not be allowed to divert attention from uglier realities, so also the perfect community must not be idealised. Well organised, representative community organisations do not emerge from the complex terrain of local politics simply because planners state it should be so. Many of the more successful examples of community based initiatives have involved outside support, including but not only from NGOs. In any case, one of the more important advantages that well organised communities have is an ability to go beyond local initiatives when necessary, and negotiate effectively with private sector and public sector actors.

Synergies Between the Public, Private and Voluntary Sectors

More generally, there is a growing recognition that when the activities of the state, the private sector and the voluntary sector can be made complementary, or even merged, considerable synergies result. Arguing over the advantages and disadvantages of plans, markets or voluntary organisations, may actually divert attention from the opportunities for achieving such synergies. Perhaps more attention needs to be paid to how the sectors interact, and to providing mutually supportive environments, and less to fighting for one sector over another. This sort of approach is at least implicit in the contemporary jargon of "enabling policies," "public-private partnerships" and the like. "Co-production" has been used to refer to the creation of goods with contributions from private and public sectors, and sanitary improvement has been cited as a good far more efficiently co-produced than produced within one sector alone. Thus for example, the condominial system which has proved to be successful in spreading comparatively low cost sanitation in several cities of Brazil, works through combining the centralised provision of trunk lines with active local involvement in financing, maintaining and even designing the connections to people's homes.

There is no guarantee, however, that cross-sectoral partnerships will create positive synergies. For example, a situation where a public utility controls the low priced public water supply, but private vendors redistribute the scarce water, lends itself to a pernicious form of private-public partnership. Public sector employees may seek private gain through restricting water supplies and entering into partnerships with the private sector vendors to capture some of the rent. Indeed, many of the same circumstances which prevent the public sector from providing services to low income settlements (including governance problems, low public sector wages, public utility financing shortfalls, illegal land settlement, and politically powerless residents), also threaten the viability of public-private partnerships.

Poverty Alleviation as a Means of Environmental Health Improvement

Ultimately, poverty alleviation would seem to be the ideal means to solve water-related health problems in low income urban settlements. Certainly it is a solution that few would openly object to. And once people are clearly willing to pay for adequate water and sanitation themselves, there are numerous comparatively successful organisational models to choose from, involving varying degrees of public and private control. Moreover, many of the water-related health problems are compounded by other aspects of poverty, such as malnutrition. Why not concentrate efforts on creating better economic opportunities, and let people buy their way out of the environmental poverty?

Unfortunately, economic development cannot be relied on to eliminate poverty any more than it can be relied on to solve environmental problems directly. Indeed, poverty, like environmental degradation, is a symptom of unbalanced economic development rather than a technical incapacity. The poor do not lack healthy water systems only because they cannot afford them, but also because they lack the local political space to organise, and the political leverage to make the public sector respond to their needs. Moreover, illness is not just a result of poverty, it contributes to poverty. In any case, it is precisely because neither free markets nor central planning readily address the problems of water, sanitation and ill health, that so much more needs to be done. Were it simply a question of getting prices right or designing the right technical interventions, water and sanitation would not be a serious problem in the first place. Giving up because there is no straightforward solution is entirely inappropriate.

Despite all the complexities and obstacles to improvement, better water and sanitation has been and will probably continue to be one of the more tractable problems which beset low income neighbourhoods. There is still a great deal for both experts and residents to learn about water-related illness. If a concerted effort is made to create and share such knowledge, health should improve considerably. In a great many cities, the organisational approach to water and sanitary improvement could be greatly improved.

It may be difficult to generalise about what an improved organisation would look like (except perhaps to make the obvious point that it must tap local collective initiative, as well as governmental and private sector enterprise). But that does not mean that improvements are difficult to devise. Indeed, the misguided search for simple and replicable solutions has itself helped ensure that there are still many opportunities for locally tailored improvements. Moreover, while problems of politics and governance undoubtedly plague efforts to improve water-related health problems, water and sanitation improvement is far less politically contentious than most other avenues for empowering low income residents.

Overall, water-related health problems are closely correlated with, but not reducible to poverty. These health problems also persist because they remain poorly understood, the appropriate institutions to address them are lacking, the political as well as economic demands of those affected are often ineffectual, and, when capital intensive solutions are absent, the physical characteristics of water related disease transmission creates a range of externalities and public goods problems. There are a number of relatively simple measures which deserve continued promotion: among residents this could include oral rehydration therapy and more handwashing; among public utilities this could include removing water subsidies to relatively affluent consumers and minimising connection costs in low income areas. But there is also a need for a more holistic approach to urban water and health, through which progress could be made on numerous fronts simultaneously. To some extent, water related health issues are already on the contemporary environmental agenda: they appeared quite prominently in Agenda 21, for example. By and large, however, in the more formal arenas water-related health problems in low income areas are perceived as a residual problem, that do not involve the complexity and uncertainty, or the institutional challenge of the new sustainability issues. This perception is incorrect.

Water and Urban Sustainability

Global urbanisation has been one of the defining features of this century, and as we move into the next century close to half of the world's residents are urbanites. A sharp distinction between urban and rural settlement has become increasingly inappropriate. Blaming urbanisation for burdens which just happen to originate in urban areas is even less appropriate. Nevertheless, for water, as for other resources, urban growth is a critical parameter, and the unprecedented urban growth of the twentieth century is a matter of some concern. At very least, this growth has helped define the water challenge of the 21st century.

Urban Growth and the Industrial Revolution

Before the 19th century the level of urbanisation had remained at just below 10% for several hundred years. The percentage of population living in urban areas in Europe and the Americas was around 75% in 1994. Urban growth and higher levels of urbanisation started with industrialisation in Europe and the Americas, closely tracking economic growth until the Second World War, by which point the initial urban transition was largely complete in much of Europe and North America. The slowdown in urbanisation rates may reflect a natural tendency for share variables to change more rapidly in the middle of a transition (the point of inflexion) than towards the end, when the upper limit (theoretically 100%) is being approached. Or, it may reflect a change of attitude toward the city, to which the new trend toward deurbanisation in some countries, for example in the United States, can be attributed.

Technological change favouring urban activities is still the most obvious explanation for contemporary urbanisation, as it was in the past. The rural-urban divide in wealth, wages and access to services remains significant. Migrants generally do better economically by moving to an urban area. As during the European urban transition, there has been a great deal of concern about over-urbanisation. Efforts to curb urban growth have been notably unsuccessful, however, and using over-urbanisation as an excuse for not dealing with urban problems has been counterproductive .

Overall, the urban growth rates in Third World countries have been spectacularly high: 4% per year in 1950-1955; over 5% in 1955-1960; and between 3.5 and 4% up to the 1990s. For cities such as Lagos, Nigeria, and Dhaka, Bangladesh, average annual growth is still beyond 5%. Starting from already high numbers, this involves some 400,000 more people in each of these cities every year.

In addition to unbalanced technical change, much of the urban 'explosion' in Third World countries derives from high overall population growth, coinciding with the middle of the urban transition. Infant mortality rates have fallen quickly (from above 200 per 1000 births to around 100) over a few decades in the middle of this century. This process took over a hundred years in the now "developed" world, where it also took a hundred years for birth rates to respond to declining death rates. However, in the Third World today, there is also a tendency toward declining birth rates.

Urban growth rates are also declining, even in Africa, the least urbanised but most rapidly urbanising continent, with a 1990-1995 rate of 4.4%, expected to become closer to 3% by 2020.

Urban Growth and Shortfalls in Infrastructure

With the current world urban population growth of 2.5%, there is an addition of over 60 million urban dwellers each year. Most of this growth is occurring in poor regions that have the least means for providing infrastructure and acceptable housing. Even though shantytowns may have been around for as long as cities have, their prevalence can be spurred by rapid urban growth. A United Nations World Housing Survey of sixty-seven large cities in the 1970s concluded that the average proportion of the population living in shantytowns was 44%. In spite of problems with definitions and fragmentary data it is clear that too many urban dwellers today live with substandard housing conditions.

High urban growth *in combination* with poor economic development can create enormous problems, but problems with poor housing should not be automatically attributed to urban growth, as pointed out by Hardoy and Satterthwaite :

> "Since 1900, the population of the Los Angeles - Long Beach urban agglomeration has grown more rapidly than that of metropolitan Calcutta. Today, both have close to 10 million inhabitants. But no one would suggest that the scale and nature of housing problems in the two centres is comparable".

More and Larger Cities

Apart from unprecedented urban growth during this century, there has also been an appreciable rise in the upper limit on city size and a marked increase in the number of very great cities. In 1950, only New York and London had a population of 8 million or more (defined by the United Nations as 'mega-cities'), but two decades later 11 cities had become mega-cities. Many excessively spectacular projections about future proliferation and size of mega-cities have been made, and are now being scaled down. Hardoy and Satterthwaite provide an example: "Mexico City's projected population for 2000 was 31.6 million in the [United Nations] Population Division's 1973-5 assessment but down to 25.8 million in the 1984-5 assessment". In The 1994 Revision of World Urbanization Prospects, Mexico City's size in the year 2000 is estimated at 16.4 million. Of course, while 16 million only represents half of the estimate provided in the 1970s, it is still an awesome size.

Close to 14% of the urban population in 1990 lived in cities of more than 5 million people. However, over 50% of the urbanites lived in urban centres of less than half a million people. As regards environmental problems, there is no reason to believe that these will be less prevalent in smaller cities, particularly if one admits inadequate household water and sanitation to be environmental problems. In most Third World nations, smaller urban centres are likely to have a much lower proportion of their populations served by piped water and sewage systems, and even less effective pollution control and land use planning, than larger cities. Furthermore, it does not require a heavy concentration of industry to bring about serious pollution problems in a small locality.

The Regional Resource Burden of Large Cities

Large urban agglomerations do create special challenges for achieving sustainable urban living. With the sheer size of the city, its capacity for generating its own resources diminishes, and water and other resources may have to be drawn from far away. Low density urban sprawl can be especially wasteful in terms of land use, and augments infrastructure costs as well as energy use and air pollution. Water supply systems become increasingly large and complex, in particular where metropolises are created out of smaller cities with originally independent water systems. And the larger the city, the larger the total generation of waste and pollution - all the more difficult for the local ecosystem to absorb.

While the increasing size of cities may further cement the urban linear material flows, to determine the environmentally most appropriate size of a city is as fruitless as attempting to pinpoint the economically ideal city size. A city's ability to grow in harmony with the surrounding ecosystem depends on the natural setting of that particular city. More generally, it is primarily the wasteful practices of the urban-industrial age that strain the world's resources, and not urbanisation in itself. With existing technologies and institutions, it can be easier to collect and treat wastes carefully when people and activities are concentrated in urban centres. Thus, the total impact on the environment might be less in a city than if the same people were dispersed throughout the countryside (with the same consumptive life-styles). However, existing urban development patterns should not be defended on the grounds that existing patterns of rural development are equally burdensome. It is necessary to address the urban deficiencies directly.

Displacing the Burden

Cities and their citizens cannot sustain themselves by drawing only on the resources within the city boundaries, and with increasing water use, sources further and further afield are often tapped. Similarly, the wastes cannot be absorbed within the city, but are displaced to the surrounding ecosystem. In the quest for new clean water and ways of disposing wastes, cities may appropriate both historical and spatial hinterlands, and through over-exploitation or pollution, undermine ecosystems and people's livelihoods far away in space and into the future. The capacity to displace environmental burdens increases with wealth and the development of centralised water supply and sanitary systems. In poor neighbourhoods, inadequate access to water and sanitation has primarily local effects, while integrated urban infrastructure allows better off segments of the population to draw on a larger resource base, protect themselves from exposure, and displace hazardous or unpleasant pollutants. Middle-income mega-cities often have particularly severe impacts on the regional ecosystem, but the wealthier, often much cleaner, cities in the 'North,' may still impose an enormous 'ecological footprint' on global hinterlands. As the environmental transition becomes 'complete,' many of the effects become globalised, contributing for example to climate change, which may alter hydrological cycles differentially across the world. As problems of insufficient water quantities and water pollution are transferred from the local to the broader environment, the challenge shifts from one of maintaining human health to one of preserving the integrity of life support systems for future generations.

The Invisibility of Linear Material Flows

Circular material flows cannot be kept within the confines of a city, where most of the ground is paved and food production can only take place on a very limited scale. Circular flows extending beyond the bounds of the city are more feasible, but not at all characteristic of existing urban development patterns. The risks associated with linear flows are rarely immediately evident, however. Indeed, the material flows characterising cities of the post-industrial era are to a large extent invisible to the cities' inhabitants. The overuse of resources and overloading of waste sinks often take place at a considerable distance from the city, and even there, may not become evident until a number of irreversible ecosystem damages have occurred. Many of the most resource and waste intensive cities are superficially getting cleaner.

Compounding this visibility problem, tracking the physical pathways of environmental impacts provides at best a partial picture of the environmental effects of a given urban demand. Markets and other mediating institutions also determine the resource repercussions of a given change in demand. In one set of circumstances, for example, by reducing demands being met by a local reservoir it may be possible to reduce pressure on distant water bodies. Thus, if there is an overall decline in demand, the reductions may occur at the more costly and distant sources. In another set of circumstances, efforts to reduce the demand on local sources may shift demand, rather than savings, to the more distant sources. Thus, if overall demand remains constant, savings from local sources may be compensated by increased consumption of the more distant sources. What scenario actually applies depends in part on the physical characteristics of the water system, but even more on the institutional context.

Urban Encroachment on Life Support Systems

Most of today's cities are located on prime agricultural land or near valuable ecosystems, so even though the estimated land surface dedicated to urban uses is only 1% of the Earth's total surface, urban expansion affects the Earth's most productive ecosystems. Such losses are serious in nations with limited arable land. Urban expansion also threatens marine and lacustrine ecosystems: 42% of the world's cities with more than one million inhabitants are located along coastlines, and many of them have been experiencing unprecedented growth. There are strong economic pressures to develop coastal areas, because of the attraction of shoreline locations, and coastal wetlands have been a particular locus of conversion. Apart from land conversion, nearby urban markets may induce over-exploitation of certain fish stocks or other resources, as well as outright destruction of natural habitats.

Cities interact with their hinterlands (and hinterlands interact with their cities - their markets) in a number of ways - many of them highly complementary. However, while natural resources tend to be drawn in to the city, the flow of 'wastes' tend to lead *from* the city along different pathways and towards destinations far from their sources. Water, being a major metabolic medium of the urban system, with 60 to 100 times the rate of the flow of fuel is an extremely effective means of environmental displacement, but one which virtually precludes the recirculation of, for example, nutrients.

History of Using Water to Displace Environmental Problems

Cities have a long tradition of using rivers, lakes and coastal waters as receptacles for diluting and dispersing wastes. The river Aire, flowing through Leeds in England, was described in the 1840s as follows:

> "It was full of refuse from water closets, cesspools, privies, common drains, dung-hill drainings, infirmary refuse, wastes from slaughter houses, chemical soap, gas, dye-houses, and manufactures, coloured by blue and black dye, pig manure, old urine wash; there were dead animals, vegetable substances and occasionally a decomposed human body".

The introduction of water-borne sewerage systems, whereby increasingly large quantities of sewage were poured into water bodies, initially without any type of treatment, tended to increase the problems of surface water pollution. At the end of the 19th century, some fifty years after London had embarked on its great sewerage scheme, the Bazalgette's sewers were daily pouring close to 700 thousand cubic meters of sewage into the river Thames, constituting about one-sixth of the total volume of the river water. The seeming lack of concern for river pollution may be attributed to the larger concern for the still very high death rates in most Northern cities in those times. This is illustrated by a statement in 1898 by Edwin Chadwick, perhaps the most famous sanitary reformer:

> "Admitting the expediency of avoiding the pollution, it is nevertheless proved to be of almost inappreciable magnitude in comparison with the ill-health occasioned by the constant retention of pollution [i.e. excrement] in the most densely-peopled districts".

This view was supported by the fact that with the extension of the sewerage system, death rates *were* declining, in spite of the increased river pollution. Still, the problems of river pollution were debated, and during the last decades of the past century "a conviction gradually developed in [the British] Parliament and among sanitary and environmental reformers that the prevention of river pollution should be... 'an indispensable requisite of every system of sewage disposal which can lay claim to efficiency'".

The awareness of the need to protect the ambient water quality was not generated by concerns for the health of the public, but for the threats severe pollution imposed on the life support systems of the cities. The minutes of evidence of the Royal Commission on the Prevention of River Pollution from 1866, are full of complaints from professional and amateur fishermen that their livelihood or sport was endangered. To impose controls of pollution in those times, as today, raised serious concerns that it might lead to severe economic dislocation. Some steps were taken to form river boards, placing river pollution control in other hands than those of local bodies, but legislation was ineffective, giving precedence to industrial interests over those of public health and conservation. National legislation, accepted standards of water quality, and larger scale sewage treatment would not be implemented until the next century, and the quality of the Thames' water would not be substantially improved until the 1970s.

> The metabolism metaphor has been used to describe cities as parasites which damage their 'host' - the environment. The concept, a variation to an input - output model where resources, through activities, are turned to residuals, can give guidance towards enhancing the complementarity of the city and its host, making the relationship more symbiotic and less parasitic.
>
> Girardet notes that while London's sewage problem was solved by building an extensive drainage system, to keep Britain's farm land fertile, guano (bird droppings) were shipped over the Atlantic from Chile.

The Role of Centralised Systems in Displacement

If each household or neighbourhood has to deal with its own human waste problems, the faecal material is unlikely to travel very far (except if there is a convenient stream or river). Faecal pollution burdens are commonly displaced from the home environment through water closets and articulated waterborne sewerage systems, and from the city environment through the lengthening of sewage outfalls or disposing of sewage sludge into the sea or landfills, or to the atmosphere through incineration. Similarly, localised water scarcity within the city may be displaced through the extension of piped water supply systems. As the central authority assumes the responsibility for providing all citizens with adequate amounts of water, to 'fill the pipes' may entail importing water from faraway systems. While dam construction and water diversions allow urban expansion and stable water use with less regard to hydrological fluctuations, river flows are reduced, and water shortages (or pollution burdens) are transferred to downstream riverine habitats.

Where the extension of the centralised systems is limited, so too is the displacement. In many low and middle income cities, public water and sewerage systems only serve the more affluent segments of the population, and hence, in many poor neighbourhoods, pollution burdens as well as water scarcity are retained locally.

Localised Urban Water Scarcities

Localised water scarcity is problematic in many squatter areas and other localities where groups of people are excluded from the use of public water systems, and lack the means to develop private facilities. For example, low income areas in Nairobi, Kenya, consume some 35 percent of the domestic supply, but account for almost two thirds of the population. The city's per capita consumption is about 90 litres per day, but only 20 litres per day in low income areas (and over 200 litres in high income areas). In the case of Guayaquil, Ecuador, the average production and supply capacity of the existing facilities would allow each inhabitant an average daily consumption of 220 litres, but daily consumption ranges from an average of 307 litres per inhabitant in the well-to-do parts of the city to less than 25 litres for those supplied by private water sellers.

While the differences in consumption levels often vary between different geographical areas, reflecting the 'limit' of the piped network, public, private and 'natural' systems can work alongside each other, but still without relieving the poor of their water problems. In Jakarta, where less than 20% of the households have piped water, most households use wells as their main drinking water source, and over 20% buy drinking water from vendors. However, the quality of the groundwater differs across the city, and in some areas high salinity makes this water unfit for drinking. Wealthy households in these areas are more frequently connected to the piped public water supply, while the poor usually buy their drinking water from vendors. Hence, a majority of the richer households will pay the lower official price, while most of the poor pay a ten times higher price per litre from ambulating vendors.

Poor households in saline areas pay on average almost 4 Rupiah per litre, while the richest households in saline areas on average only pay 2 Rupiah per litre. In areas with "potable" groundwater, however, the poor can resort to well water which is not paid for, and hence pay less than the average richer person, who still is more likely to have a piped water connection and pay the official price for water. These price differentials reflect the fact that the urban poor are more dependent upon the residual natural water resources of their own localities, while the wealthy have come to depend upon a man-made system that spreads the burdens over a far broader area.

The situation in Jakarta reflects a somewhat extreme example of a common tendency for the poor to rely on "secondary" water markets for drinking water, being unable to access the municipal water distribution system directly. That in many parts of the world the poorest segments of the urban population suffer from chronic water scarcity does not just reflect their lack of market "command." There is ample proof that poor city dwellers often pay more per litre of drinking water than the richer urbanites (Bhatia and Falkenmark, 1993; Kjellén *et al.*, 1996; Swyngedouw, 1995; World Bank, 1988). Often, the principal difficulty is to get a water connection, not to pay for the water.

From Health to Sustainability Concerns

Poor sanitary environments take their toll on health rather than on the ecological foundations of the city. Faecal-oral diseases are endemic in densely populated settlements that lack adequate water supply and sanitation. In better serviced neighbourhoods, ample amounts of wash-water help reduce transmission, and pathogens, most of them in human excreta, are flushed away through a public sewerage system. Where treatment is lacking, pathogens may well appear in ambient waterways, but are less likely to be encountered, especially by households from the polluting areas. In some cases polluted ambient water is ingested, directly or by eating seafood, especially uncooked 'filter-feeders.' There may also be adverse health effects where wastewaters are used for irrigating crops. However, the main impact of urban sewage is on the health of aquatic life, and not public health.

The release of pollutants from urban areas threatens ecosystem integrity in numerous ways. Alterations in the levels of dissolved oxygen and nutrients, salinity and acidity change the environmental conditions to which the aquatic life are acclimated to. While it is difficult to identify and assess specific causal relationships that threaten biotic communities, species are lost to extinction, and impoverishment of biotic diversity can lead to less stable and less productive habitats. Adverse ecological effects may also affect economic productivity in the short run.

An Extreme Example of the Displacement of Water Burdens

Mexico City has evolved as a mega-city in a water scarce region essentially through the 'successful' displacement of both water pollution and water shortage. The mining of groundwater and large water transfers allow a per capita use of over 300 litres per day, sustaining forty-five percent of the country's industry and serving ninety-four percent of the city's population of over 15 million people. The alteration of the natural hydrological system was a prerequisite to the city's expansion. The Valley of Mexico, originally a closed basin, was artificially opened in the 1700s in order to control flooding. This has also enabled Mexico City to remove its effluents out of the basin. Without this, or some alternative form of displacement, Mexico City's aquifer would have been destroyed by pollution long ago. Still, poverty prevails, and the city has yet to supply all its people with water: 6% of the population - almost a million people - have neither household connections nor neighbourhood standpipes. These are likely to suffer from water scarcity, facing greater health risks, but imposing less of a water resource burden. And even among the connected households, average water consumption levels are far less than those of Mexico's Northerly neighbours.

Wealth Creation and the Appropriation of Hinterlands

Because of their ability to innovate, humans are not bound by a 'carrying capacity' defined by existing technology. However, a major component of humankind's 'ingenuity' has been to draw on an increasingly extensive resource base, and displace waste to increasingly distant locations or diffuse it more broadly. In other words: "From the beginning of the European empires of the sixteenth century to the industrial empires of the present, the richer people in the richer countries have appropriated *global* hinterlands to meet the needs of their increasingly urban citizens". These developments have accelerated during the last fifty years. As put by Rees, "...the five-fold increase in the scale of human economic activity in the post-war period has begun to induce ecological change on a global scale which simply cannot be ignored in planning for human settlements". Larger cities' pull on resources, in these days through trade, may be one of the major historical changes happening to planet Earth. The urban-industrial society is displacing resources such as soil nutrients or fossil carbon into the sea and to the atmosphere, and mixing some of the more hazardous wastes into toxic 'cocktails.'

Empirical Observations on Pollution and Affluence

Multi-national studies support the view that city-wide pollution problems increase initially with economic development (as more pollution is generated and also displaced from the home and community), but subsequently decrease with further economic wealth. The level of dissolved oxygen in rivers tended to decrease in the worst off rivers in low-income countries in the 1980s. In China, only five of fifteen river stretches near large cities sampled by UNEP's Global Environmental Monitoring System in the mid-1980s were capable of supporting fish. During the same period the level of dissolved oxygen increased in the cleanest rivers in high-income countries. Phosphorous levels have declined markedly in many Western European rivers from the end of the 1970s, primarily because of intensified wastewater treatment, but also the substitution of phosphorous in detergents. Wealthier cities tend to have more stringent environmental legislation and the institutional means to ensure compliance, as well as more resources to spend on environmental quality. But even in the richer North, river pollution problems are frequently 'displaced' rather than 'solved.' Sometimes sewage outfalls are only extended in order to avoid local pollution effects, and even where wastewaters are treated, the sewage sludge may be dumped at sea.

Clean City with a History of Pollution

The cleanest cities of our times are among the wealthiest, with good quality water supplied to both industry and households and most effluents treated in order not to foul the local or the regional environment. In these cities the epidemiological transition away from water-related infectious diseases has been virtually completed, and persistent environmental problems tend to have poorly understood, indirect or delayed impacts on human health. Many currently very 'clean' cities in the developed world have for long periods of time been releasing polluted water, thus accounting for far more than their share of the build-up of contaminants in the oceans.

Stockholm, Sweden, situated on the boundary between the lake Mälaren and the Baltic Sea, proudly dubs itself as one of the cleanest cities in the world, at least when it comes to the water environment. Still, the city has passed through a trajectory, maybe less critical, but still similar to that described above for London. For much of the last century, the mortality rate was higher than the birth rate, at least in part as a result of water related diseases. Then the quality of local ambient waters went from bad to worse with the increased use of water closets, and by the 1930s the waters were considered unhealthy and bathing houses were closed down. Treatment of wastewaters started in 1934, and by the 1970s treatment had been extended to having all wastewater mechanically, chemically and biologically treated before being released into the sea. Lake Mälaren, the city's water source, was also regulated to avoid brackish sea-water and (treated) wastewaters to enter from the sea, and upstream sources impairing the quality of lake waters were addressed through co-operation with other municipalities around the lake. As the emission of phosphorous and oxygen-demanding substances have been considerably reduced, plankton production in the inner archipelago has decreased and water has become clearer. Trout have been restocked, and annual swimming competitions, suspended in the 1920s, were reassumed in 1976.

Stockholm has been successful in addressing the regional impacts on its water system and its immediate water surroundings are remarkably clean. Also, over half of the sewage sludge produced is used as agricultural fertiliser, but problems remain relating to the acceptability of this practice. While Stockholm may not qualify as a 'hotspot' of obvious pollution to be addressed along the coasts of the Baltic, emission of nitrogen has remained unchanged and still favours plankton production in the outer archipelago. Much 'historical pollution' remains in sediments. For the Baltic Sea, the more large-scale problems, relating to long-range transport of pollutants through both water and air, exist in the open sea, to which historically Stockholm may be a major contributor.

Indirect Routes of Undermining Water Systems

With cleaner cities and city surroundings, the major challenge is again shifted, now towards the more far-reaching threats of the urban-industrial systems to global ecology. High energy consumption, based largely on fossil fuel, is to a large extent an attribute of wealthy cities in the 'North' - especially those with colder climates requiring extensive heating. This contributes to global climate change whose predicted effects include sea-level rise and increased variability of local hydrological cycles.

The purchasing power of urban elites and the masses of middle-class urban consumers to a large extent dictates which commodities are produced, and even the production processes. As regards diets, urban dwellers generally eat more meat and water-intensive crops. To produce one kilo of wheat requires some 500 litres of water, a kilo of rice, up to 2000 litres, and some 20-50 thousand litres of water per kilogram of meat. To produce one automobile consumes (directly and indirectly) close to 400 cubic meters of water. Increasing numbers of more environmentally aware and concerned citizens prefer goods that are produced in ways that are less damaging to the ecosystem, but such preferences are not easy to translate into effective demands.

Water, Wealth and Sustainability

Different Perspectives on Sustainable Development

Although a city and its citizens cannot be self-sustaining, urban systems should be capable of developing production and consumption patterns that do not unduly compromise the local, regional or global ecological balance. Unfortunately, unless one interprets "unduly compromise" very loosely, this is not a simple achievement. From an ecocentric perspective, urban expansion almost inevitably compromises sustainability by reducing ecological system efficiency and integrity at numerous levels. But even from an anthropocentric perspective, urban development typically threatens sustainability, not least through its effects on water systems.

A major principle of sustainability is not to compromise the future; for economists not to deplete resource stocks and for ecologists to preserve biodiversity and not to exceede critical threshold levels. There are however a great variety of views regarding the place of humans in a sustainable world, as well as the confidence in technological innovation and resource substitution as means of avoiding Malthusian scarcity.

On the anthropocentric side, the 'Brundtland Report' saw the main urban challenge as that of service provision in cities in developing countries, and was quite optimistic about the future of urban centres in industrial countries: "The combination of advanced technology, stronger national economies, and a developed institutional infrastructure give resilience and the potential for continuing recovery to cities in the industrial world. With flexibility, space for manoeuvre, and innovation by local leadership, the issue for industrial countries is ultimately one of political and social choice. Developing countries are not in the same situation. They have a major urban crisis on their hands".

From a different perspective, however, while the local and regional problems in many less affluent cities may indeed be acute, the 'creeping crises' of the resource base depletion, global warming, ozone depletion and acidification is the major threat to global sustainability. Affluent city dwellers are more responsible for these global threats, which have resulted from the high aggregate use of non-renewable resources and waste accumulation. From an ecological viewpoint "humankind remains a creature of the ecosphere existing in a state of obligate dependency upon many products and processes of nature" and intergenerational equity would be expressed, not in terms of productive stocks of capital, but rather: "Each generation should inherit an adequate stock of natural assets *alone* no less than the stock of such assets inherited by the previous generation". Urbanites, while physically and psychologically distanced from the ecosystems that sustain them, appropriate carrying capacity through trade. From this perspective, the urban impact on sustainability has been globalised.

Population, Affluence and Technology

No single factor (urbanisation, population growth, technology, ever increasing affluence in the Western world, etc.) can be singled out as the root cause for humanity's increasing threat to global sustainability. Ehrlich et. al. portray the human impact through a 'web of blame,' where the environmental impact equals *population x consumption of goods per person x environmental impact per quantity of goods consumed,* abbreviated as "impact equals population times affluence times technology". As relations are multiplicative they reinforce each other; an environmentally disruptive technology is thus more damaging in a large rich population. However, the factors are not independent, with both demographic patterns and technologies differing systematically between rich and poor societies.

Moreover, at every level good environmental management makes a difference. Effective removal of health threatening and 'poverty related' environmental hazards are made easier by more resources in the hands of society, but are just as contingent upon effective governmental action and public health policies. As regards environmental problems of regional or city-wide importance, where emission levels have eventually come to decline with rising incomes, environmental legislation, institutional reforms and incentives to reduce environmental impacts have been the proximate motivation for improvement. Hopefully, the same will one day be said of the global issues.

The global sustainability outlook depends to a large extent on the evolution of aggregate resource consumption levels. Regarding overuse of global sinks, equity concerns may be less critical from a physical point of view, and are not captured in aggregated models and figures, but will be of utmost importance in forming amelioration strategies. It is tempting to apply the principle of subsidiarity, and allow local institutions to deal with local environmental problems, city governments to deal with city level problems, and for global mechanisms to be devised to address with global threats. However, this would be neither fair nor effective. It would be unfair inasmuch as it gives governing institutions the responsibility for coping with the problems of affluence, but not those of poverty. It would be ineffective inasmuch as local initiatives are also critical for global problems, and, as argued in a previous section, local initiatives need international support whether they are globally or locally oriented.

From Displacement to Replacement

Many of the regional water burdens of cities can be alleviated by improved collection and treatment of wastewaters, and the use of sewage sludge as agricultural fertiliser. This implies improvements to existing systems following the convention of using water as the means for transporting wastes, but often a re-orientation of sewage works to function as 'fertilizer factories rather than as disposal systems'. Recycling in this sense would entail 'curving' the current 'open ended' linear flows, increasing 'exchange' with the rural hinterlands (farmers), rather than a 'closed loop system.' Challenges for improvement relate to keeping sludge clean from toxic pollutants, distance to farming communities, city growth, climate and the availability of freshwater. Traditional (pre-industrial) systems of waste water disposal may provide useful models for more natural, ecologically compatible systems, costing much less than modern high technology. Purification of wastewaters may be technically possible, but still not economically viable even for the world's richest cities. The fixed standard (capital intensive) sewage treatment model of the North is increasingly challenged in favour of approaches with flexible (and more participatory) standards, balancing the trade-offs between cost and water quality.

To be successful, such changes will require new institutional forms and that new interest groups capable of providing new technologies be built up. Existing, centralised, waste disposal systems are very path dependent: once a particular approach has been selected, strong forces can come into play ensuring that this approach be extended rather than replaced. The technologies are readily available, and even the education system becomes oriented towards their use. To change this path will undoubtedly take considerable effort, and is likely to be uneconomic in the short run.

Refining Dilution

A more refined, and less damaging, way to 'displace and dilute' can be achieved by adapting to and not exceeding critical thresholds in terms of damaging the natural ecosystem. This entails making maximum sustainable use of nature, or, inversely, induce maximum survivable damage. Water bodies' vulnerability to contamination depends upon the volume of flow which determines how effectively the pollutants are dispersed; large rivers are less vulnerable than small rivers and lakes tend to be more susceptible to contamination than the seas. Following an approach of 'sensible dilution,' uniform water quality standards may not be strict enough to protect some inshore waters, such as lagoons, from irreversible pollution damage, but overly strict for other better-flushed waters. Nevertheless, effluent standards tend to be determined on the basis of broad political goals rather than on the water quality and self-purification capacity of the receiving water.

To apply this approach extensively in a centralised fashion would require both a profound knowledge of local conditions, and very sophisticated and incorruptible government regulations. On the other hand, it is not yet clear what sort of participatory standards are feasible. Again, there is likely to be a need for new institutional forms.

Conservation Through Higher Prices

Partly as a reaction to the call for closed cycles, *reducing* the linear flows is increasingly seen as a potential avenue for modifying and ameliorating the effects of the presently mainly linear systems. Curbing water use (and waste in particular) may be addressed by demand management, where 'management' is a euphemism for reduction. The volume of the water flow through cities with well developed infrastructure is essentially determined by the demands. The immense throughputs, especially of the wealthier cities in the world, can be reduced through awareness campaigns, price signals as well as water saving technologies.

The use of price signals is gaining increasing international support. That "water has an economic value in all its competing uses and should be recognised as an economic good" has been adopted as the guiding principle for water management at several international meetings, notably at the International Conference on Water and the Environment: Development Issues for the 21st Century, held in Dublin 1992, and is actively promoted by international agencies such as the World Bank. Given the economic value of different uses, such water allocation is in most cases likely to favour urban water use over agricultural uses, but could also mean that environmental uses are compromised unless nature's needs can be effectively monetised.

Problems with Externalities and Water Pricing

For price signals to guide actors towards sustainable water use, they need to reflect the full environmental cost - the intended or unintended environmental effects of polluting or abstracting the water. The price would reflect the value of the foregone opportunity for other potential users of the same water, or in the case of pollution fees, foregone opportunities because of degraded quality. While damages compromising societal needs are difficult to correctly appreciate in monetary terms, damages to natural environments that are not exploited can be far harder. When dealing with resources like water, efficient (economic) allocation is inherently problematic for a number of reasons:

> "Water's vital and often revered role has led many societies to restrict selling water and pricing it to reflect its full cost and scarcity. But even in the absence of societal reservations about treating water like other goods, the nature of the resource makes it difficult and in many cases impossible to establish efficient markets... Efficient markets must satisfy two conditions. First, there must be well-defined and transferable property rights... Second, a market transfer is efficient only if the full benefits and costs are borne by the buyers and sellers. Both conditions are likely to be violated for water resources".

Market allocation of water needs efficient institutions to avoid distortions ranging from monopolistic practices and impacts on third parties to protecting the environment and safeguarding equitable allocation for basic human needs. For example, in the absence of mechanisms for protecting common property resources, farmers in Northern California, USA, sold surface water rights to California's State Water Bank, but replaced the sold water by pumping (unregulated) groundwater.

Marketing of pollution or abstraction rights, giving opportunities for environmental groups to purchase rights (without using them), can put price tags on environmental values, but can be very difficult to manage, and ignores important differences on the demand side. First, it is not at all clear that the economic burden of protecting the environment should fall primarily on the environmentally aware. If saving water is a public good, then it is highly inappropriate to expect environmentalists to voluntarily purchase water protection on the market. Alternatively, in an area where water use is necessary for public health, it is also inappropriate to have people paying full resource costs.

Price based demand management approaches make most sense in situations where different users, or on a more aggregate level - different uses, are competing for the same water. The most obvious rivalry for water resources on the larger scale is the competition between agricultural and urban water abstraction, but the (often unintentional) use of water for dispersing pollutants may be equally challenging within and around cities. Basin wide approaches are needed to defend downstream users from undue appropriation of water by upstream users, and, in particular, to defend environmental needs.

Ecological Cities and Barriers to Change

While current urban practices can be improved in numerous ways, adjustments to existing systems will fall far short of any real change of the material foundations of the city. It is argued that "only a radical break with modernity can overcome the multidimensional crisis characterizing our epoch and hasten the necessary restructuring of the built environment of our cities and its relation to the wider natural ecology". Whether an 'ecological city' is achievable or not can be debated, as can the meaning of 'ecological' in the urban context. These uncertainties make any concerted action towards a fixed goal impossible. For water, its many different functions in society, conflicts of interests, as well as insufficient knowledge about many of the environmental effects, adds to complexity.

Barriers to substantive changes include the already (expensively) built-up environment in the developed world as well as conceptual, institutional and social barriers towards changing existing systems. Stricter pollution controls have setbacks in terms of economic dislocation - raising concerns of 'deindustrialisation' in many developed countries. Relatively little hope is given to 'technocrats' to solve fundamental problems - many of the present urban environmental challenges (such as traffic congestion) are the *unintended* outcome of *intentional* decisions. While citizens' scope for action is to a large extent hindered by being a small part of a much larger system, the push for system change may need to come from active concerned citizens. Urban areas are often held out as potential centres of origin for substantive changes:

> "The ecological challenge with which industrial society is confronted is thus not only a question of technology, but above all a question of lifestyles and social values. In this respect we need innovations and ecologically sound solutions to problems in civilization. History has shown that such solutions may be found, given the will to survive, just because of the innovative power of cities".

While social values are important for legitimising and reinforcing urban systems and technologies, they do not necessarily support ecological restructuring. (Top down) environmental education may be a prerequisite for 'grass root' city dwellers to actively pursue more ecologically sustainable lifestyles. Furthermore, well educated citizens can be much more effective in voicing their concerns, and pushing for broader policy changes. The major benefit, however, of relying on local initiatives, is that it is at the local level that lifestyle changes need to be made, and taken alone tools like pricing and regulation are in danger of seeming oppressive, and are in any case not likely to be sufficient for the task. Increased democratisation, through participation of city dwellers in the decentralised planning may be the only viable way for real change, regardless of whether people are 'reduced to mere consumers' or are seen as pro-active agents.

There is still a serious institutional problem to be addressed, however, if cities are to take a lead role in the pursuit of sustainability. If one city concentrates its efforts on cleaner streets and another city on reducing it contribution to global warming, one city ends up with cleaner streets and both end up in a slightly more sustainable world. Unless some higher incentive is given to the city that attempts to address global warming, a city-based effort is almost certain to stall. The search for "win-win" opportunities is useful in its way, but cannot provide the basis for a long term strategy.

Water for Healthy and Sustainable Cities

Cities are to a large extent the product of selfish pursuits. People settle, work, trade, and invest in the city primarily to serve their own ends, rather than those of the city as a whole. Often these self serving pursuits inadvertently contribute to the dynamism and attraction of the city. But they can also unintentionally contribute to environmental problems, undermining public health and sustainability. One of the age-old roles of urban government has been to help manage the urban environment, including its water systems, and to prevent some of the excesses of unregulated development. With the rapid urban growth of the last two centuries, urban health and sustainability have emerged as critical international concerns, and motivated concerted actions within civil societies as well as governments. The sanitary movement of the 19th century and the environmental movement of the 20th century have both been urban-based, and gathered considerable public support. Unfortunately, much remains to be accomplished, with regard to both health and sustainability, and not least in relation to urban water systems.

Unity Through Diversity

Looking across today's variegated world, environmental awareness clearly requires a different focus in different cities. A city's water strategy and politics need to evolve around the most pressing needs for that particular city. And, needless to say, but overwhelmingly difficult to implement, a water strategy should address the needs of all citizens. Since living conditions in many cities vary radically, this can entail parallel strategies targeting different issues for different water users in different parts of town. Thus in a single city it may be appropriate to target conservation for households in one area, network extension and increased water use in another area, and to develop different policies with regard to informal and formal sector enterprises. Such policy variation is superficially inconsistent, but only if one adopts the simplistic view that everyone is entitled to inexpensive water, or that everyone should have to pay the full economic cost for their water. Actually, economic efficiency alone would dictate a very different approach in areas where public health is at risk due to water scarcity, from areas where, for example, neighbours are competing to have the greenest lawn. In short, a standardised approach to urban water management will be divisive while an approach that serves everyone needs to§ diversified.

Superficially, emphasising the distinction between health and sustainability concerns might seem to add to sectoral divisions and policy fragmentation. However, to achieve coherence, as well as balance, differences as well as similarities need to be recognised. Perhaps because they are both environmental problems, while environmental health and sustainability involve quite distinct threats, they are structurally rather similar. Threats to both sustainability and health are principally the unintended outcome of human pursuits. Often the threats are hard to perceive. Considerable headway can often be made through holistic responses, centred on abstract and somewhat vague concepts such as *cleanliness* with respect to health and *natural* with respect to sustainability. Both pose a challenge to science inasmuch as the processes remain poorly understood, and to governance inasmuch as the environmental interconnections create a host of what economists term externalities.

Local Action Based on Local Realities

As indicated in previous sections, it is not surprising that health concerns dominated during the 19th century. In many of the world's most affluent and dynamic cities - the motors of the industrial revolution - environmental health conditions were appalling. Currently, sustainability concerns dominate, at least in the international environmental arena. Again, this is not surprising. In the world's more affluent cities, water related diseases are no longer a major concern; the more obvious challenge is whether these cities' relatively newly found water affluence can be sustained. However, if the urban environmental agenda is to gain grass-roots support internationally, it will need to take the local environmental problems, and contemporary urban health concerns, back on board.

As the environmentalist cliché "think globally, act locally" suggests, sustainability issues have a multiplicity of scales. The city scale has been regaining prominence, but not principally because the damage caused by urban centres is better recognised. For water, as for many other environmental concerns, urban areas do concentrate damaging activities. But more important, for there to be a transition to more sustainable development, urban areas must also become concentrations of environmental improvers. Local *initiatives* are increasingly seen as central to achieving sustainability. And it is from this perspective that the pursuit of sustainability needs to be grounded in local urban realities.

One urban reality, however, is that the immediate health and welfare aspects of water, rather than the longer term sustainability aspects, remain the more pressing environmental concerns for a great many of the less affluent urban dwellers. If cities and urban neighbourhoods are to become the "grass roots" of environmental reform, then the environmental agenda must encompass the more pressing "grass roots" concerns. However environmentally aware, a low-income peri-urban dweller lacking piped water and sanitary facilities is likely to be more concerned with water and sanitation than with water and sustainability. The importance of local initiatives is increasingly recognised in the context of environmental health improvement in low income settlements. This greatly reinforces the potential for creating a more broad based urban environmental movement, encompassing both health and sustainability concerns.

International Cooperation

Despite the need for locally tailored solutions, international cooperation is important. This applies to both health and sustainability. The need for international cooperation is self evident as regards sustainability, where international water bodies and cycles are often threatened. But it also applies to environmental health improvement, where the hazards are predominantly local. The sanitary movement was unequivocally international, despite the local nature of the environmental problems addressed. Its success was predicated on the fact that reformers from different countries shared science, practical experience and inspiration. In today's world of jet-hopping and global conferencing, there is no dearth of opportunities for international exchange. Unfortunately, these opportunities have not been capitalised upon to create a coherent approach to urban environmental improvement. More specifically, the urban environmental agenda remains fragmented.

For most developing countries, international development cooperation is of course as important as the international exchange of ideas and experience. Still, external financial assistance poses yet another challenge to good governance and local participation. The international influence on local water systems is not surprising, given the pivotal role of international financing of water development in many Third World cities. Indeed, it is critical that donors make every effort to ensure that aid funds are not misappropriated. And there is an inevitable tension between the interests of local residents and the interests of local governments and government officials: interests that correspond in official rhetoric, North and South, but rarely in practice. This tension does not, however, justify supporting projects and policies which ignore the expressed desires of both local citizens and their governments, to conform to an externally defined environmental agenda. Rather it makes the job of defining a politically appropriate water strategy that much harder.

Challenges for Governance

The pursuit of both environmental health and sustainability also poses challenges to good governance because the beneficiaries are not usually well represented in government. Groups with serious water-related health problems typically live in conditions of poverty, and are politically as well as economically disadvantaged. Groups severely affected by a water-related threat to sustainability include especially future generations. Representing the interests of the poor poses quite different political challenges from representing the interests of future generations. On the other hand, the justification is similar, and there is at least one strong political reason for addressing these two sets of interests simultaneously: it is critical that representing the interests of future generations not be used as an excuse for sacrificing the poor, and vice versa. Both of these groups are almost inevitably included in political rhetoric, but easy to ignore in policy implementation.

Orthodox water policies do not provide a sound basis for pursuing either environmental health or sustainability, at least in part because these most affected groups do not have sufficient voice. The process of defining the goals and principles of water policy faces the same risk of being hijacked by more influential segments and entrenched groups as any other political process, especially where democracy is fragile or nonexistent. Even where democracy is secured, the interests of future generations are not thereby represented. On the other hand, the potential for making headway is considerable, since there is very little public opposition to the pursuit of public health and sustainability. This makes well designed policies, participatory mechanisms, and good science especially important, as they provide the means to create urban environment strategies that are enforceable in addition to being well intentioned.

From Productionist Logic to Demand Management

In practice, piped water and sewerage systems in developing countries often fail to reach the neediest parts of the population, but still create considerable city-wide environmental burdens. When water supplies are expanded in disadvantaged areas, residents are rarely given the hygiene education needed to ensure that they can meet their own health needs effectively. Alternatively, water companies find themselves lagging further and further behind in supplying ever increasing city populations with clean water. They run structural deficits and often operate with ad hoc interventions as and when additional funding becomes available. According to the World Bank, in most developed countries consumers pay water rates covering all recurrent costs and a large share of capital costs, while in developing countries water prices cover only a third of the average cost of supply.

The politically determined tariffs and the negative returns on water sales, the historical preoccupation with massive engineering structures for the production and transmission of water, and the bias towards providing unlimited quantities of water to industry and well-to-do households; these all contribute to both chronic water shortages and the systemic exclusion of large parts of the population from access to the available water. Based on the notion that the water requirements are given and will be met only if the supply system expands, it is as if a 'productionist logic' directs water companies towards new investments and consequent dependence on external finance and its inherent technological bias. This has led to many gigantic engineering projects, made possible in the 1970s through ample access to international loans. For many countries, this has contributed to an enormous foreign debt. It has increased access to water for millions of people, but both economic and water resources have been squandered in the process. And it has opened the door to reactionary water policies which address the economic and resource inefficiencies, and can claim to do no worse, and perhaps even a little better, with respect to the currently unserved.

Demand management questions the basis of 'productionist logic,' treats the volume and pattern of water use by individuals, households, businesses, farmers, etc. as variable, and aims to change the behaviour of consumers either voluntarily (prices, education, etc.) or involuntarily (regulation or centralised interventions). Where supply ends and demand starts in the water distribution system is debatable, but a common denominator for demand management, as it is currently used for water management, is to encourage use efficiency and water conservation - at all levels. The tendency is to emphasise water conservation. However, in areas where access to water is insufficient, and water scarcity is creating health problems, demand management should involve increasing supplies accompanied by hygiene education. More generally, while demand management is typically viewed as a tool for achieving more sustainable water systems, an alternative form of demand management is central to the more sophisticated attempts at improving environmental health. Thus, just as demand-side conservation is increasingly seen as critical to water resource management, demand-side hygiene behaviour is increasingly seen as critical to water related health improvement.

Involuntary methods, such as rationing, have been a frequent response to water shortages, especially short-term and severe ones. However, interruptions to water services, intentional or not, have several adverse effects. Apart from lowering the pressure in the system (increasing the risk of cross-contamination from wastewater), necessitating additional investments (household storage cisterns and sometimes booster pumps), and disrupting both economic and social activities, it is doubtful how effective it actually is in curbing consumption. In anticipation of water cuts, households may tap more water than is actually required during the time of service interruption, water of which large proportions are subsequently wasted. Also, where all households are filling (automatically or manually) their cisterns during that part of day service is provided, peak demand may be higher than what it would be given a continuous service. Moreover, water interruptions pose a variety of health risks, and have been found to be statistically associated with a higher prevalence of diarrheal disease. In many circumstances, providing a continuous rather than intermittent supply of water can save both water and money.

Higher water prices can be an efficient way of curbing aggregate demand (as long as metering, billing and revenue collection actually take place). While low volume users are less likely to respond to price changes, high volume users (with declining marginal utility), such as industries, commercial establishments and also affluent households should reduce water consumption in response to higher water prices. In areas with dangerously low water consumption, pricing and other components of water policies should not target conservation *per se*, although in many circumstances better demand-side management will lead to water savings even in disadvantaged neighbourhoods.

Targeting Wastage Rather than Usage

More generally appropriate than reducing consumption is the goal of reducing waste by preventing leakage from the supply systems. How much water leaks out of a system is rarely known; it typically falls under the general category of *unaccounted for water,* which, strictly speaking, includes all water which is not metered. Ranging from unintentional (technical) leaks from the pipe system, to water distributed to official non-paying customers, such as for fire-fighting or sometimes military camps, unaccounted for water in many cities hovers around half of the quantities supplied. A lot of the unaccounted for water is of course put to beneficial use, even though all of it presents a severe challenge for utility companies' financial sustainability. Leaks, however, with the exception of groundwater recharge, have no benefits. Rather, they contribute to accumulation of stagnant water, disrupting transport as well as providing breeding grounds for many disease bearing vectors. Furthermore, they can create negative pressure in the water pipes, allowing sewerage to enter the water system. Even in the short term reducing leakage can lead to significant savings, and simultaneously provide health benefits.

Affordability and Efficiency - Beyond the Water Tariff

The main argument for keeping low water prices to urban households is that poor urban dwellers can not afford to pay the full cost of meeting basic human water needs. Although there is evidence of exorbitant water prices in many poor areas, these prices typically often only reflect what is paid for the highest value uses; drinking and cooking. Water for washing and cleaning would rarely support such high prices, and such uses may be severely compromised if all water prices are uniformly high. Furthermore, evidence of high willingness to pay for water must not overlook the social cost of other consumption opportunities foregone, nor the personal deprivation and public cost of insufficient water for hygiene.

To overcome affordability problems, water tariffs often allow a lower unit price for small consumers; a "lifeline rate" with a unit price that would be augmented with higher levels of consumption. Given the nature of water distribution in many poor areas, where existing facilities tend to be shared by many users, the desired effect may however not materialise even for those with water connections. A household selling water to its neighbours will have a higher meter reading, and may, in spite of servicing a higher number of people, pay more per unit of water. While a progressive tariff may serve to dissuade some wealthy households from using water carelessly, it can thus also be a burden on households who share water meters or sell water.

Furthermore, official pricing policies may only be directly relevant for those households that receive their water directly from the utility. Where water distribution networks only serve parts of the urban area, subsidised "lifeline" rates are highly unlikely to benefit more than a small minority of the poorer urban dwellers. And worse, where universally low water prices render utility companies unable to expand the network to under-serviced areas, the exclusion of the poor is made permanent. With constrained supply leading to high price differentials, the end result may even be that while the water consumption of the wealthy is subsidised, the poor are effectively being charged more than what the cost would be, given an economically efficient distribution system. In the extreme, charging higher prices for water, at least to affluent users, could benefit the poor, and also encourage conservation among businesses as well as households using municipal water for gardening and washing cars. However, there is no guarantee that a more financially viable utility will extend services to low income areas, even if water prices are set to cover costs.

Effective demand management, for both sustainability and health, requires going beyond water pricing. To pursue water sustainability in an effective manner there are a wide variety of possible measures, many of which have other environmental benefits. Indeed, it has been estimated that the 1992 U.S. Energy Policy Act, which mandates water efficiency standards for all residential faucets, shower heads and flush toilets, could eventually result in a decrease in all domestic water uses of 20%. With respect to health, demand side management can also involve a variety of measures, ranging from cross subsidising household connection costs to community consultation in designing payment systems to the promotion of selected technologies and hygiene behaviours.

The Importance of Being Well Connected

There is a great deal more to water affordability than the official water price, and for many households the connection cost alone matters more than the water price. The economic justification for subsidising the connection cost for low income households is more compelling than that for water itself. Health studies indicate that having a water source in the home does make an appreciable difference. Subsidising new connections in deprived neighbourhoods can help target the incentives, and provide a greater incentive than equivalent subsidies on water itself. As the success of institutions like the Grameen Bank indicates, low income households have difficulties raising capital. They often face the equivalent of annual interest rates of several hundred percent. Thus, for poor households, making it easy and inexpensive to obtain a water connection can be economically much more important than equivalent financial resources devoted to keeping the price of water low.

Alternatively, when water prices rise, or household incomes fall, it is possible to use demand management to provide alternatives to disconnection. Even in the U.S. there is a significant population of low income households for whom water bills seem unaffordable. To avoid disconnection, and the bad publicity this creates, some utilities are practising demand management, and exploring options such as financial counselling, arrearage forgiveness, payment discounts, income-based payments, lifeline rates, targeted conservation, disconnection moratoria and flow restriction. Yet such measures, in a somewhat different form, are even more important to low income cities, where, as both recent riots and epidemiology duly demonstrate, disconnection is not just a private affair.

Redefining the Actors

Virtually all large scale urban piped water systems have been developed through public channels. Still, it is only the 'transformation of nature's water' and bringing it to the city which is typically a wholly public activity. The 'urbanisation of water' is in many countries organised through a mixture of public and private systems. In developing countries, while more affluent households and powerful enterprise tend to be served through a wholly public system, poor households more typically rely on private systems often reselling water from the public system.

A holistic approach to urban water - from abstraction to return to the water cycle - needs to encompass all the actors of the water's pathway through the city. This should include the often large numbers of households that do not have house connections. Where individual connections may not be possible, utilities should have approaches to deal with groups as well as with individual households, requiring yet more differentiation of tariffs. Alternatively, developing intermediary institutions, which can help ensure that utilities' practices meet local community needs, can considerably lessen the gap between centrally developed policies and local concerns.

As mentioned above, the state, the private sector and the voluntary sector can be made complementary, and much attention, especially within development projects, has been brought to citizens' groups' roles in water management. Collective action within neighbourhoods can do a lot to resolve sanitary problems. However, the relative efficiency of many community organisations should be no excuse for governmental inertia, partly because effective neighbourhood solutions are often contingent upon the prosperous relations and support from governmental agencies.

There is no Easy Agenda

To provide households with water supply and sanitation services has been referred to as an 'easy task,' but recognising the complexities regarding unequal income distribution, the externalities involved and problems of democratic representation of the poor, as well as the enormous backlog in achieving targets, this is presumptuous. There is a need to recognise the complexities and aim at understanding them better, and how they relate to other problems and priorities city dwellers and urban governments have to deal with.

All implementation is in a sense testing of old or new concepts under new circumstances. Some 'success stories' such as the Orangi project in Pakistan or condominial sewerage systems in Brazil are frequently described in international publications. Learn from these examples about old and new technologies or how systems of finance can be successfully combined, or how much can be achieved by individual charismatic leaders with social commitment and the right support from governmental agencies. But underestimating the level of skills, commitment and ingenuity that can bring about success in this difficult area may be counter-productive. As the specific nature of the problems vary between cities and parts of cities, any 'standard' solution has to be adapted rather than replicated. Furthermore, a lack of humility to the task tends to put inordinate blame on those daring to take action, and discourages prominent leaders and professionals from pursuing endeavours in the area.

No Excuses, but Humility to the Task

The challenge to make our cities more healthy and more sustainable is indeed enormous. Poor sanitary environments at this very moment continue to bring death, disease and human deprivation, and water pollution and overuse is increasingly threatening our life support systems, regionally and globally. The need for action is certainly there. Disciplinary and institutional boundaries must be transcended. Local knowledge and scientific knowledge must be brought together into well targeted and well coordinated efforts. However, the policies for good coordination, as the knowledge base for well informed strategies, are still in the making. Underestimating the complexity of the task, regardless of whether the challenge is to improve health or sustainability, may lead to costly mistakes. Our ignorance is no excuse for inaction: there are many obvious things that need to be done. But nor is the need for action an excuse for remaining ignorant: there remains a great deal of relevance to learn regarding water, health and sustainability.

REFERENCES

Covich, A. P. (1993). Water and ecosystems. In: Gleick P.H., (ed.). *Water in Crisis: A Guide to the World's Fresh Water Resources.* New York, Oxford University Press. pp. 40-55.

Dilks, D., (ed.) (1996). *Measuring Urban Sustainability: Canadian Indicators Workshop,* Toronto, 19-21 June, 1995. Ottawa, Ontario, Canadian Housing Information Centre. 73 p.

Douglas, I. (1983). *The Urban Environment.* London, Edward Arnold. 229 p.

Frederick, K. D. (1993). *Balancing Water Demands with Supplies: The Role of Management in a World of Increasing Scarcity.* World Bank Technical Paper No. 189. World Bank (Washington, DC). 72 p.

Girardet, H. (1990). The metabolism of cities. In: Cadman D., Payne G., (eds.). *The Living City: Towards a Sustainable Future.* London, Routledge. p 170-180.

Hahn, E. (1991). *Ecological Urban Restructuring: Theoretical Foundation and Concept for Action* No. FS II 91-402. Science Center Berlin (Berlin). 37 p.

Hamlin, C. (1990). *A Science of Impurity: Water Analysis in Nineteenth Century Britain.* Bristol, Adam Hilger. 342 p.

6

Impact of Climate Change on Freshwater Ecosystems

Freshwater ecosystems have been critical to sustaining life and establishing civilizations throughout history. Humans rely on freshwater systems not only for drinking water, but also for agriculture, transportation, energy production, industrial processes, waste disposal, and the extraction of fish and other products. As a result of this dependence, human settlements worldwide are concentrated near freshwater ecosystems, with over half of the world's population living within 20 km of a permanent river.

In addition to humans, an enormous array of plants, animals, and microorganisms depend on freshwater ecosystems for their survival. Although freshwater ecosystems contain only 0.01% of the Earth's water and cover a small fraction of the planet's surface, rivers, lakes and wetlands harbor a disproportionately high fraction of the Earth's biodiversity. Freshwater fishes alone account for over one-fourth of all living vertebrate species..

Because freshwater ecosystems continuously channel precipitation from the surrounding landscape through the interconnected lakes, rivers, and wetlands that lie below, they can be surprisingly sensitive to distant activities. Increasing human water needs and extensive land alteration has contributed to the decline of countless freshwater species. Freshwater biodiversity is now more threatened than terrestrial biodiversity, and the projected mean future extinction rate of North American freshwater animals is about five times higher than for terrestrial animals, and comparable to predicted extinction rates for tropical rainforest communities.

The long-term protection of freshwater species is largely dependent upon identifying the underlying physical processes of freshwater systems that are most vulnerable to change, and determining how changes in these physical features might affect the resident flora and fauna. For this purpose, it is useful to divide freshwater ecosystems into rivers, lakes, and wetlands. In this chapter, rivers and streams are considered to be channelized bodies of water that generally display continuous flow, and lakes are relatively still bodies of water that can be either connected (through rivers, streams, etc.) or isolated from other bodies of water. Wetlands (also known as marshes, swamps, fens, bogs, floodplains, or depressions) are areas where the water table is at or near the surface, and vegetation is submerged for at least part of the year. Many, although not all, wetlands are connected to or interact strongly with lakes and rivers.

Because freshwater ecosystems are sensitive not only to water temperature, volume, and flow, but also to variability in these factors, rivers, lakes and wetlands are expected to display a wide variety of changes in response to global climate change.

Threats to Freshwater Ecosystems

Because freshwater ecosystems depend strongly on physical features such as water quantity, quality and flow, many of the threats to these ecosystems involve activities that alter fundamental physical characteristics. Freshwater ecosystems throughout the world are threatened by human activities that directly alter system hydrology, such as construction of physical barriers to flow, water extraction, and filling or draining of shallow habitats. Pollution of waterways with toxic substances and excessive nutrients, as well as destructive land use practices in areas surrounding freshwater ecosystems, lead to reductions in water quality. While the above threats directly affect physical features of freshwater ecosystems, the introduction of exotic species primarily affects native biota. The invasion of freshwater ecosystems by non-native species is rapidly becoming one of the most serious threats to freshwater communities. Overexploitation of animals associated with freshwater ecosystems, particularly freshwater fishes, is also a continuing problem. Finally, penetration of harmful UV-B radiation into water bodies is increasing in many areas due to interactions between a number of anthropogenic factors, and a range of negative impacts on freshwater communities may result.

Alteration of Hydrology Physical Barriers to Flow

Humans have constructed a variety of physical barriers, including dams, levees and dikes, to prevent flooding, generate power, supply water for irrigation or municipal water supplies, and provide recreational opportunities. Dams have been built on every continent except Antarctica; they are prevalent in developed countries and their rate of construction is increasing rapidly in developing nations. Dams have traditionally been viewed as an environmentally-friendly and sustainable means of ensuring water supply, controlling floods, and generating power without polluting the environment. However, retaining water and altering its natural flow can lead to large changes in aquatic and terrestrial habitats, both above and below dams.

Initial flooding above dams to create reservoirs can result in massive losses of terrestrial habitat; for example, India lost approximately 479,000 ha of forest land to various river valley projects from 1950-1975. Stagnation and low flow rates in reservoirs can lead to large changes in water temperature, including variations in seasonal peak temperatures and a reduction in natural temperature variation. Silt that is normally carried down rivers accumulates behind dams, and costly removal procedures are sometimes necessary to ensure that dams remain functional. In addition, the decomposition of flooded vegetation above dams may release significant quantities methane and other greenhouse gases, particularly in tropical areas where plant biomass is high ; the magnitude of this greenhouse gas release compared to fossil fuel burning has been widely debated.

Dams also cause dramatic changes in downstream flow regime, where seasonal, dynamic flows are replaced by steady water release for energy production, or intermittent large releases to lower reservoir levels. The temperature variability of water released from the bottom layer of reservoirs is low compared to that of natural stream flows, and oxygen content may be reduced. The loss of natural silt in released water can lead to a range of damaging effects downstream, including changes in chemical composition, river bank erosion, and massive habitat loss and erosion in coastal deltas and floodplains.

Finally, dams pose a significant barrier to diadromous fish and other migrating animals. In the Brazilian Amazon, where many fish undergo long-distance migrations during the rainy season, 79 dams are either planned or currently in existence, and hydroelectric development is considered to be the greatest threat to Amazonian fisheries in the near future. On the Columbia River in the northwestern United States, 19 major dams have already contributed to the extinction of at least 106 local stocks of pacific salmon.

Humans also build obstructions such as levees and dikes to prevent water from flowing laterally over river banks during high flows—in this way, land adjacent to rivers can be developed with less risk of seasonal flooding. However, these barriers disrupt connections between rivers and valuable floodplain habitats, which serve as refugia and spawning grounds for many animals, and are often sites of high biodiversity.

Water Diversion/Withdrawal

In order to meet the agricultural and municipal water needs of a growing population, large quantities of water are diverted or withdrawn directly from rivers, lakes, or the underlying water table. A large proportion of the world's population is currently experiencing water stress, and human water needs are expected to increase dramatically in the coming decades due to projected population growth and increased development.

Withdrawal of water for human needs can reduce the total amount of water available to aquatic biota (resulting in low stream flows and declining lake levels), reduce seasonal variability in flows, and lead to large losses of habitat, particularly valuable edge habitats that are used for spawning or rearing. Withdrawal of groundwater for agriculture in adjacent areas can completely dry up valuable wetlands, even if the habitat itself is protected. Water that is withdrawn from freshwater systems is often returned in the form of household wastewater, agricultural runoff, or industrial cooling water, with reduced quality (increased pollutants, changes in nutrient load) and altered temperature.

Filling or Draining Shallow Habitats

Wetlands throughout the world have been filled or drained for development or agriculture. These habitats have traditionally been regarded as useless in their natural state, and many countries have provided subsidies to encourage the conversion of wetlands to agricultural land. Thankfully, the valuable ecosystem services that wetlands provide, including water purification, groundwater recharge, and flood control, are becoming more widely appreciated.

Invasive Species

Apart from direct alterations of system hydrology, the most serious threat to freshwater ecosystems in many areas is the presence of invasive, non-native species. Invaders are often adaptable generalists who breed and disperse quickly, endangering native species through highly efficient competition, predation, or habitat alteration.

Deliberate Introductions

The deliberate introduction of non-native species for commercial or recreational fishing is widespread; for example, approximately 80% of the alpine lakes in the western United States have been stocked with non-native fish. In many areas, lakes and streams are continuously re-stocked to maintain populations. Although these fisheries generate enormous income, the effects on freshwater communities can be devastating. Hundreds of native fish species in North America are threatened, and competition or predation by non-natives is the primary threat to many of these species. In Lake Victoria, increased predation and competition due to the introduction of the Nile perch and several non-native tilapiine species (in addition to ongoing environmental changes in the lake), led to the extinction of up to 200 endemic cichlid species within a few decades.

Accidental Introductions

Many species are introduced to and spread throughout freshwater ecosystems acciden-tally—for example, non-native species can be attached to boats or transported in ballast or bilge water. The zebra mussel, which is native to Europe, was accidentally introduced to the Great Lakes in 1986 through ballast water. It has altered water chemistry through highly efficient filter feeding, outcompeted populations of native mussels, and fouled boats, docks, and power plant intakes, costing millions of dollars in damage. Several species of native mussels, clams, and commercially important fish are threatened directly or indirectly by its presence.

Non-native plants can also cause damage to aquatic ecosystems. The water hyacinth, which is native to the Amazon River, has been introduced to freshwater ecosystems on several continents, including Africa and North

America. It grows on the surface of water, blocking light, decreasing oxygen levels, and changing water chemistry. Entire food webs have been altered as a result, and fish populations have been reduced or eliminated in some areas. In the United States, water hyacinth has spread to rivers, lakes and lagoons throughout the country, forming dense mats that block canals and drainage pipes, prevent swimming and boating, and impair waterway navigation. In Lake Victoria, water hyacinth "hot spots" have persisted near areas of urban, industrial, and agricultural pollution, despite efforts to eradicate the floating vegetation.

Pollution

Freshwater ecosystems are polluted by a variety of human activities, from large-scale agriculture and industry to everyday behaviors, such as driving cars and fertilizing lawns. Large quantities of pollution often enter freshwater systems from point sources, such as industrial or municipal sewage outflows; for example, 23.4 billion tons of sewage and industrial waste was dumped into the Yangtze River in 2001, threatening human health and the survival of the endangered Yangtze River dolphin. Thus, the focus on protecting water quality in many countries has been on preventing point source pollution. However, nonpoint source pollution is far more significant in many cases. Airborne pollutants can enter the atmosphere and travel long distances, entering lakes and waterways in otherwise pristine locations. Pollutants dissolved in runoff from the surrounding landscape may account for the greatest source of pollution in many freshwater ecosystems—for example, it has been estimated that 80% of the nutrients (nitrogen and phosphorus) that pollute U.S. waterways derive from nonpoint sources such as agricultural and urban runoff.

Nutrient Pollution

Runoff from fertilizers used in commercial agriculture or private yards adds large amounts of nitrogen and phosphorus to freshwater ecosystems. This can be especially problematic in lowland areas and in lakes or rivers with developed shores. The added nutrients lead to excess growth of algae (which is sometimes toxic), resulting in reduced water clarity and light penetration. Because of this increased primary productivity, the activity of decomposing, oxygen-consuming bacteria increases and oxygen levels decline. Shifts in the food web and alterations in bottom-water habitat can lead to changes in species composition and distribution. For example, the density, distribution and relative abundance of aquatic plants can change after eutrophication, and valuable fish species are often replaced by less desirable fauna that can tolerate low oxy-gen levels. Natural eutrophication is a normal state in the succession of lakes as they age, but polluted runoff has led to early eutrophication and changes in the community structure of many naturally oligotrophic (nutrient-poor) lakes.

Toxic Pollution

Toxic pollution in freshwater ecosystems can devastate local biota and endanger human food sources. Most toxic pollution derives from industry (i.e. dioxin, PCBs) or agriculture (pesticides such as DDT and toxaphene). Heavy metals, such as arsenic, zinc, selenium, and mercury are also released from mining and other industrial activities. Heavy metals and toxic compounds can become volatized in warmer parts of the globe, enter the atmosphere, and re-condense in cooler areas, often contaminating pristine sites and indigenous food supplies. These pollutants cause massive mortality events, such as fish kills, and are found in concentrations considered unsafe for human consumption in many aquatic animals, such as freshwater mussels. Because many pollutants accumulate in fatty tissues, they are magnified in the food chain and can reach concentrations in fish-eating birds and mammals that are up to 10 million times higher than those found in polluted water.

Acid rain containing high levels of sulfuric and nitric acids is also a serious threat to many freshwater ecosystems, particularly lakes at high altitudes and latitudes. These pollutants enter the air mainly through the burning of fossil fuels, are transported atmospherically to distant parts of the globe, and are finally released into freshwater ecosystems through precipitation. In lakes susceptible to acidification (where acids are not readily neutralized by the soil or water), lake pH is lowered dramatically, and species composition and abundance can change as a result.

Land Use

Destructive land use practices that result in vegetation loss anywhere within the drainage basin of a river can have negative impacts on freshwater ecosystems. The forests and native plant communities surrounding lakes, rivers, and wetlands help protect water quality and quantity by filtering and storing runoff. Changes in land use brought about by agriculture and urbanization (such as deforestation, chemical fertilization, and paving) lead to increased runoff with higher levels of nutrients and other pollutants. In addition, more sediments are washed into the system and water turbidity rises, with negative impacts on fish, filter feeders, aquatic plants, and bacteria.

The vegetation immediately surrounding water bodies (i.e. riparian or floodplain vegetation) is particularly important to freshwater ecosystem health, as it shades water bodies (regulating water temperatures and providing thermal refugia) and supplies organic material such as falling leaves, insects, and woody debris to freshwater systems. The deforestation of Amazonian floodplains is considered to be one of the major forces behind the decline of Amazonian fisheries, as deforestation in these seasonally flooded areas has led to massive erosion, increased sediment load, and a decrease in large, woody debris in the river. Although its function has not been studied extensively in tropical rivers, large, woody debris has been shown to play numerous important roles in temperate river systems, including altering flow to create habitat heterogeneity (i.e. stepped channels and deep pools) and helping to determine river channel form and stability. Woody debris also has a long residence time in most water bodies, and provides substrate, food, and shelter for a wide range of plants and animals.

Overexploitation

Fish and shellfish harvests have declined sharply in the last few decades, and in many cases this decline is due to commercial or recreational overexploitation of these resources. Overharvesting of aquatic species has occurred repeatedly in freshwater ecosystems throughout the world, in part because high levels of natural population variability can mask the effects of overexploitation until population declines are severe and irreversible. Attempts to bolster declining populations, such as with the release of hatchery fish, often exacerbate the problem. Species inhabiting lakes can be particularly vulnerable to overex-ploitation in areas where there is not continual recruitment from populations outside the lake.

Hunting or extermination of mammals associated with freshwater ecosystems (i.e. beavers, muskrats) can affect not only the biological community, but also the physical structure of freshwater ecosystems, as the activities of some species lead to flooding and the creation of important wetland habitats. Hunting in tropical rivers and floodplains can add additional pressure to endangered species such as manatees, turtles, river otters and caiman.

Exposure to Ultraviolet-b Radiation

Exposure to high levels of Ultraviolet-B (UV-B) radiation can have a range of harmful effects on living organisms, and damage to the ozone layer has caused an increase in UV-B radiation of up to 50% in some alpine areas. Although the ozone layer protects both terrestrial and aquatic ecosystems from much of the sun's UV-B radiation, some aquatic animals may be more vulnerable to the radiation that reaches the Earth's surface because they have historically been afforded a high level of protection by natural characteristics of the waters they inhabit. UV-B radiation generally only penetrates the top layer of water bodies; the depth of UV-B penetration depends on water clarity and on the concentrations of dissolved organic carbon (DOC) and/ or chromophoric dissolved organic matter (CDOM), which are derived from the breakdown of plant materials or dead organisms.

Animals living in clear or shallow bodies of water, those that are sessile or can't sense UV-B radiation, and those that are restricted to the upper layer of water during the day-time (i.e. small zooplankton taking refuge from downward migrations of predators) may be particularly vulnerable to increases in UV-B radiation. Primary productivity has decreased as a result of increased UV-B radiation in some cases, but productivity could also increase, depending on whether the system is nutrient or grazer-limited. Many vertebrates, including amphibians, lay transparent eggs with little UV-B protection in bodies of water. UV-B radiation can reduce hatching success, retard larval growth rates, and cause morphological abnormalities in amphibians, and increased exposure to UV-B has been implicated as a possible factor in (although not the sole cause of) recent worldwide declines in amphibian populations. However, the danger of UV-B radiation in freshwater ecosystems remains site-specific; for example, 85% of the potential amphibian breeding sites sampled in the U.S. Pacific Northwest are protected by naturally-occurring levels of DOC in the water.

The threat of UV-B exposure may be magnified by other human-induced stresses. For example, CDOM derived from healthy, forested drainage basins is far more effective at attenuating UV-B than colorless DOC from other sources, and deforestation decreases the input of CDOM to freshwaters. In addition, acidification of lakes causes a decline in DOC concentrations, and thus UV-B radiation may actually be responsible for many of the negative effects attributed to lake acidification.

Observed Effects of Climate Change on Freshwater Ecosystems

Human activities within the last century have led to a dramatic rise in atmospheric concentrations of carbon dioxide and other gases that contribute to the greenhouse effect. Within the next century, carbon dioxide concentrations are expected to at rise to levels at least twice as high as those present in pre-industrial times, and global climate is expected to change in a number of ways as a result. This climate change will primarily affect freshwater ecosystems through changes in water temperature, quantity, and quality, as well as through changes in the timing and duration of flows. Some of the expected physical and biological effects of climate change will affect all freshwater ecosystems, while others are specific to rivers, lake or wetlands.

Physical Effects Common to All Freshwater Ecosystems

Temperature Changes

There is widespread consensus that the greenhouse effect will lead to a global rise in air temperature, with mean surface temperatures increasing 1.5 to 5.8 °C by the year 2100. Temperatures are expected to increase more at higher latitudes, and in many of these regions the effects of global warming have already been documented; in Canada, mean air temperatures, water temperatures, and evaporation have all increased in the past 20-30 years, and ice cover durations over lakes and rivers have decreased over the entire northern hemisphere by almost 20 days since the mid-1800's.

However, in most cases, the effects of global warming on air and water temperatures are likely to be far more complicated that a gradual increase in average temperatures. In many regions, daily minimum air temperatures have increased more than daily maximum temperatures, leading to a reduction in the diurnal temperature range. Both observational studies and models of future climate change suggest that there will be more hot summer days and fewer cold waves. Regionally, temperatures are likely to become more variable, and this increased variability (i.e. a 1 °C increase in the standard deviation of temperature) will lead to a far greater frequency of extreme temperature events than a similar change in the mean temperature would.

Rising temperatures are generally expected to lead to an increase in glacial melting, although increased winter precipitation could compensate for ice loss in some areas. Many simulations suggest that glacier melting will depend strongly on the rate of temperature change; for example, Oerlemans et al. predicted that in the absence of increased precipitation, a rise of 0.4 °C per decade would eliminate nearly all of their study glaciers by 2100, while a rise of 0.1 °C per decade would only lead to a 10-20% loss of glacier volume. Tropical glaciers may be especially sensitive to global warming, as the equilibrium line between ice accumulation and melting is more sensitive to changes in air temperature (due to the lack of seasonality in tropical temperatures), and because glacial melting is significant year-round.

Precipitation Changes

Since 1900, surface precipitation has generally increased in mid- and high-latitude areas, and decreased in the tropics and subtropics. Current models of global climate change suggest that annual precipitation is likely to increase further in high and mid-latitudes and most equatorial regions, but decrease in the subtropics. However, predicted regional changes in precipitation are far less certain, and in many areas, the size of predicted precipitation changes due to global warming are small compared to those due to natural multi-decadal variability.

However, even slight changes in average precipitation could lead to substantial increases in the variability of precipitation events; because the size of precipitation events is not normally distributed about the mean, a change in average precipitation will also cause a change in variability. Climate change models predict that global warming will generally lead to more extreme events, such as heavy 1-day and multi-day precipitation, and an increase in the frequency of extreme rainfall has been observed in the United States and the UK. In addition, most countries that have experienced a significant increase or decrease in precipitation also experienced a disproportionate change in the amount of precipitation falling during extreme precipitation events. Tropical storm intensity has increased, and is expected to increase further in some areas, such as southwest Asia, but in other areas tropical storm frequency and intensity is expected to remain the same or decline.

Water Quantity and Flow Changes

Although precipitation is one of the main factors determining water availability and flow, other factors such as evaporation, soil moisture, groundwater recharge, and glacial and snowmelt are also critical. Evaporation is generally expected to increase due to increasing mean temperatures. However, the exact amount of evaporation that will occur at a given site is determined by a host of other factors, including soil characteristics, the amount of water available, vegetation cover, and plant transpiration. Soil moisture will depend on soil characteristics and the magnitude of local precipitation changes, and soil infiltration and water-holding capacity will in turn determine the volume of run-off. For example, drier soils often shows reduced water infiltration, and less extreme freezing events can reduce water infiltration in limestone soils ; reduced water infiltration could lead to greater run-off and an increase in flooding events. Groundwater recharge is affected by both the amount of precipitation and the duration of the recharge season, as well as by evaporation and soil moisture. Although climate change is likely to lead to some changes in ground-water recharge, freshwater ecosystems that primarily receive input from groundwater are likely to experience smaller changes in water temperature and quantity than those dominated by precipitation.

Water flows depend primarily on precipitation in tropical and arid regions. In tropical river systems, seasonal heavy rainfall events already surpass the natural infiltration rates of soil, leading to high sediment input and dangerous levels of pesticide runoff from agricultural lands ; increased extreme rainfall events in these areas could lead to further water quality problems. In higher latitude regions, temperature changes will affect water flow through changes in snowmelt and the form of falling precipitation. In large parts of eastern Europe, European Russia, central Canada, and California, a major shift in streamflow from spring to winter has already been observed, because elevated temperatures cause precipitation to fall as rain rather than snow. Similarly, glacier-fed rivers, lakes and wetlands (in temperate and tropical regions) may experience increased flows due to glacial melting, even in the absence of increased precipitation.

Many models predict that extreme water flow events such as floods are likely to increase, due to heavier individual rainfall events or increased overall precipitation. Some models also predict seasonal shifts in peak flooding seasons. Finally, models of climate change suggest that hydrological droughts should increase in frequency, but this has only been observed in some areas, such as Hungary and China. The effects of climate change on low flow conditions appear to be sensitive to the storage capacity of the system; basins with little groundwater storage capacity may experience more frequent droughts because they do not benefit as much from winter groundwater recharge.

Water Quality Changes

The input of chemicals, sediments, organic matter, nutrients and pollutants to freshwater ecosystems are all likely to be affected by climate change. Both simulations and direct observations indicate that increased precipitation can increase water alkalinity via enhanced weathering and input of base cations to streams or lakes. Intense storm events following prolonged dry periods can lead to increased flushing of sediments or nitrates into water bodies.

Rising temperatures and changes in precipitation are likely to cause changes in the bio-mass, production, and composition of terrestrial communities surrounding lakes, rivers and wetlands. These changes may affect the supply of organic matter to freshwater systems, shading and light (including UV-B) penetration, as well as the characteristics of runoff entering the system (i.e. DOC and nutrient concentrations, sediment load). Rising temperatures and evaporation may cause an increase in fires in some regions. Fires in regions surrounding water bodies could lead to increased nutrient input (from burnt vegetation) leading to eutrophication, increased sediment load, reduced input of organic matter, and reduced protection from winds.

Water pollution could be affected in many ways by climate change. Increased volatilization of pesticides, PCBs, and heavy metals in warm and warming regions with condensation elsewhere will lead to increased pollutant loads in water bodies at higher latitudes and altitudes. Increased runoff could lead to increased pollution from agricultural and urban sources, while decreased water levels could lead to concentration of pollutants from point sources and the atmosphere. Increased glacial melt has been shown to increase the concentrations of pollutants in glacial-fed streams and lakes by releasing organic pesticides and PCBs deposited in glacial ice during earlier decades when pollution levels were higher.

Increasing Human Water Needs and Extraction

Rising air temperatures and evaporation are likely to contribute to increasing human water use and water shortages. By far, the largest proportion of current and predicted future water use is for agricultural irrigation; in 1995, 67% of all water withdrawals and 79% of all water consumed worldwide was used for agriculture, whereas municipal, or domestic, use represents only about 9% of withdrawals.

Based on predicted population increases and development scenarios, water withdrawals are expected to increase 23-49% over 1995 levels by the year 2025. Municipal water withdrawals may be offset by declining per capita usage in some countries, and increases in water usage by the industrial sector in developing countries (particularly in Asia and Latin America) may be partially ameliorated by greater industrial efficiency. Agricultural water usage, on the other hand, is expected to increase due to higher evaporation rates and because larger areas of land are likely to be under cultivation (to feed an growing human population). Areas where agriculture accounts for a large portion of the GNP and water is already scarce (i.e. many African countries) are likely to be hardest hit by climate-associated water shortages. Alterations in the way that water is priced (i.e. reducing agricultural water subsidies) could play an important role in promoting more efficient water use and thus decreasing human water needs and extraction.

Biological Changes

Effects on Physiology and Life History

Temperature change alone is known to affect a range of physiological processes and life history traits. Higher ambient temperatures increase the metabolic demands of many animals; for example, even at sub-lethal temperatures, warming would lead to a several-fold increase in the energy requirements of lake trout *(Salvelinus nomaycush).* The effects of higher metabolism on growth may depend on food availability. In zooplankton with an adequate food supply, increased temperatures lead to a dramatic rise in feeding, assimilation, growth, and reproductive rates, and local species richness can increase as a result. Increased temperatures can also lead to a rise in the frequency of toxic algal outbreaks, and in their toxicity to other animals.

Te mperature affects body size in many aquatic animals; increased rearing temperature causes a reduction in body size at a given developmental stage in over 90% of coldblooded, aquatic animals studied. Temperature also determines the sex of offspring in the American alligator *(Alligator mississippiensis)* and several groups of turtles. In one population of painted turtles *(Chrysemys picta),* offspring sex was shown to be highly correlated with mean July air temperatures; statistical analyses indicate that a 2 °C rise in air temperatures would drastically skew sex ratios, and a 4 °C rise would virtually eliminate all males from the population. The phenology, or timing of life history events, is also affected by ambient temperature; for example, breeding migrations and spawning dates have begun to occur earlier in several species of amphibians as a result of climate warming.

Life history traits are intricately linked with water quantity and seasonal flow in numerous aquatic animals. In tropical rivers, many fish undergo feeding and spawning migrations of several thousand kilometers that are dependent on predictable, seasonal flooding events, and extreme flow levels may be necessary for maintaining populations of a number of other species. On the other hand, intense flooding can scour streambeds, displacing organic matter, bottom-feeding organisms, and small fish fry, and substantial increases in flood frequency could cause a shift in species composition, possibly eliminating many species.

The timing and duration of the breeding season in wading birds that frequent freshwater wetlands is strongly tied to water levels, and could be af-fected by either increases or decreases in precipitation and water table levels. Frogs are particularly sensitive to decreased precipitation; low precipitation in Puerto Rico has been correlated with drastic declines in frog populations, and the extinctions of four frog species in Costa Rican cloud forests have been linked to a series of severe population declines following extreme El Niño-associated droughts. Some inhabitants of seasonal wetlands, such as fairy shrimp, are entirely dependent on precipitation-filled, ephemeral vernal pools to complete short, highly-specialized life cycles.

Effects on Community Composition and Dynamics

Climate warming is likely to alter the composition of many communities, as different species will have different thermal tolerances and interactions between species may intensify as a result of reduced resources and habitat availability. In algae, thermal tolerance can affect the outcome of competition for nutrients and alter community composition. Where water levels or the size of suitable thermal habitats decrease, biotic interactions (including human overexploitation) may intensify as a result of increased densities of aquatic animals. For example, caddisfly competition in streams that is usually limited to the summer season persisted year-round in years when drought eliminated normal winter density reductions. Climate change may affect motile vs. non-motile species differently, leading to differences in species distribution; in the marine intertidal zone, vertical distribution of marine invertebrates is tightly correlated with temperature, whereas the distribution of motile species is not.

In temperate or high-elevation tropical systems, where water temperatures are currently cool, climate warming is expected to facilitate the spread and establishment of non-natives, especially those from warmer climates. Once thermal barriers to invasion are removed, native species may be displaced by invaders with a competitive advantage, such as warmwater, omnivorous fish with fast life cycles. Natural high flows help minimize the success of non-native fish by removing species that are poorly adapted to dynamic river environments; the restriction of flooding by reservoirs has already helped facilitate the proliferation of exotic fish in many river systems. Reduced flooding due to climate change could similarly allow non-native fish to become established in areas where the current flow regime would otherwise exclude them. Several species of exotic marine invertebrates may benefit from warming temperatures at higher latitudes, as earlier recruitment in warmer years helps them gain a competitive advantage over native species; a similar breeding strategy may be present in invasive freshwater invertebrates.

Migration may be the only option for many animals that cannot adapt to increasing temperatures. During past eras of climate change, most plants and animals displayed range shifts, rather than morphological change, in response to changing environmental conditions. However, the rate of warming expected in the next 100 years is over ten times higher than warming after the last ice age, and it is unknown whether plants and animals will be able to migrate quickly enough to keep up with climate change. Meta-analyses indicate that hundreds of plant and animal populations have already shown highly significant, non-random changes in range boundaries, in the direction expected to result from climate change, and temperature is known to the affect geographic distribution of fish, diatoms, and other aquatic animals.

Unfortunately, the ability of many freshwater species to migrate is restricted by bodies of water that preclude migration or that don't allow movement in the correct direction. Migration is impossible from many isolated lakes and wetlands, and numerous major river systems run from east to west, precluding latitudinal migration. In the southwest and southern Great Plains (USA), nearly all major river systems run from east to west, and these systems contain some of the hottest free-flowing water on Earth. Many native species in these areas are already living near their thermal tolerance limits, and the combination of increased warming and the lack of northern migration routes could cause extinctions of up to 20 species of endemic fish.

Even where migration is possible, it is unlikely that entire communities and ecosystems will be transplanted intact. Species differ in their levels of tolerance of environmental change, and in their abilities to adapt or migrate in response. Differences in sensitivity to temperature change are likely to alter community dynamics. For example, communities composed of several species of *Drosophila* that were allowed to migrate between different thermal habitats did not simply shift to new zones when temperatures were increased to simulate global warming. Interactions between species and with parasites introduced to the system altered the relative abundance of species in different thermal zones in ways that were non-intuitive and difficult to predict. Thus, simple "climate mapping", or assuming that species range boundaries will shift smoothly with changes in temperature, oversimplifies the effects that climate change may have on communities.

Effects of Climate Change on Lakes

Physical Effects on Lakes

Increased mean surface temperatures are likely to lead to increased water temperatures and evaporation in many lakes, in both temperate and tropical areas. If precipitation does not increase enough to compensate, this could lead to reductions in outflow and/or lake volume. Important spawning and rearing habitat near the edges of lakes would be lost if lake levels declined, and lake characteristics based on water outflow could change dramatically. Lakes that currently supply outflow to downstream systems may become endorheic (with no outflow), and endorheic freshwater lakes may become saline. The African Great Lakes are particularly sensitive to climatic effects on outflow, as current outflow is small (i.e. only 6% of water input to Lake Tanganyika leaves as riverine outflow), and even minor declines in precipitation (10-20%) are expected to completely close these basins.

Changes in mean air temperatures have been shown to increase water temperatures in both temperate and tropical lakes, resulting in a range of physical and biological effects. Water temperatures in Lake Tanganyika, a deep tropical lake in East Africa, have risen by 0.2 °C at the lake bottom and 0.9 °C at 100 meters since 1913. Although the resulting change in the temperature gradient is relatively small, the vertical gradient in water density (which depends on temperature) has tripled. This sharpened density gradient has reduced annual mixing, which normally supplies nutrients from decomposition on the lake bottom to surface waters. The lack of nutrients in the upper layers of the lake has led to a 70% reduction in primary productivity since 1975 and an increase in water clarity and light penetration.

Temperate lakes display sharper thermal gradients and larger seasonal changes in water temperature than tropical lakes. Lakes at higher latitudes and altitudes currently experience seasonal thermal stratification, in which they are covered by ice in the winter (with cool, relatively constant temperatures below), and develop a thermal gradient (thermo-cline) in the summer as surface waters warm up. Increased ambient temperatures have led to an earlier onset of thermal stratification, longer ice-free periods, and deeper ther-moclines (thus smaller bottom layers) in many temperate lakes. In addition, longer periods of thermal stratification with little vertical mixing results in reduced oxygen concentrations near the bottom of lakes.

Lake chemistry may be affected by climate change in a number of ways. Drought and decreased groundwater flow may make some lakes more susceptible to acidification, as groundwater often contains acid-neutralizing chemicals important to lake buffering. However, the overall pH and chemical balance of lakes may be affected by temperature and precipitation changes in ways that are site-specific and difficult to predict; a number of European alpine lakes that experienced a 1 °C increase in temperature over 10 years actually displayed an increase in pH, as well as trends in sul-fate and nitrogen concentrations that were opposite to trends in atmospheric deposition of these compounds. These changes are most likely explained by increased biological activity and enhanced weathering of surrounding substrates due to high precipitation.

Concentrations of colored DOC, which derives from surrounding vegetation and provides the greatest protection from UV-B radiation, has been shown to decline in high latitude lakes as a result of climate warming, reduced streamflows, and lowered water tables. Lake acidification (which can itself be caused by climate change) will exacerbate this effect. Conversely, increased precipitation should increase DOC concentrations, and vegetation shifts due to climate warming (such as altitudinal shifts in the tree line) may protect higher lakes by increasing DOC concentrations.

Finally, the physical effects of climate change on temperate lakes can be synergistic and complex. For example, lakes at the Experimental Lakes Area in the boreal forests of northwestern Ontario experienced an average increase in air temperature of 2 °C over 20 years. This resulted in an increase in mean and maximum water temperatures, an ice-free period that was 20 days longer, and an 30% increase in evaporation. The area experienced below-average precipitation and reduced runoff into lakes. More frequent fires caused a rise in nutrient input, which combined with higher temperatures to increase phytoplankton abundance and diversity. Reduced terrestrial input to the lakes (because of fires and clear-cutting) reduced DOC concentrations and led to clearer lakes with less protection against UV-B radiation. Increased solar penetration and higher winds (due to the loss of trees) led to deeper thermoclines and a reduction in bottom habitat.

Biological Effects on Lakes

Although the effects of climate warming on the biotic communities of tropical lakes has received little attention, large decreases in primary productivity due to climate warming are likely to have a significant impact on the rest of the food chain. In temperate lakes, the radical physical changes in lake chemistry and thermal stratification resulting from climate change can have a range of effects on biological communities. As temperate freshwater fisheries play an important economic role in many countries, a great deal of research into the effects of climate change on freshwater ecosystems has focused on temperate freshwater fish.

Temperate fish can be divided into three major guilds according to their "fundamental thermal niche", or the temperature at which the choose to spend most of their time and at which they experience optimal growth, activity levels and swimming performance : coldwater fish such as salmon and trout (with an optimal temperature around 15 °C), coolwater fish such as perch (with an optimal temperature around 24 °C), and warmwater fish such as carp and catfish (with an optimal temperature around 28 °C).

Seasonal thermal stratification in lakes allows all three guilds to co-exist because of variations in life history strategies. In the winter, cool- and warmwater fish are inactive, and cold water fish are active but experience little growth. Temperatures are optimal and growth rates highest for coldwater fish in the spring and fall, while the summer is optimal for cool- and warmwater fish. During the summer, the upper layers of the lake become too hot for coldwater fish and they are restricted to the cooler bottom layer, where oxygen levels are low and competition for food is fierce.

Because climate warming leads to longer periods of thermal stratification, coldwater fish will be restricted to these bottom layers for longer periods of time, and deeper ther-moclines brought about by climate change may reduce the area of bottom layers and further intensify competition for food. Climate change may also lead to shorter spring and fall seasons, when temperatures are optimal for coldwa-ter fish. An overall rise in water temperatures will lead to increased metabolic demands, but coldwater fish will generally have reduced access to prey. Warmer temperatures may make winter slightly more favorable, but not enough to compensate for losses during other seasons. Overall, coldwater fish are likely to experience decreased growth rates and increased heat mortality.

Some populations may be able to adapt to thermal changes; a shift in preferred temperature from 15 °C to 20-21 °C has been observed in one population of lake trout (Salveli-nus namaycush) in Canada, and other populations of coldwater fish display adaptive behavior, in which they spend most of their time in coldwater refugia but make occasional feeding forays into warmer water. However, the majority of coldwater fish populations are likely to experience range shifts, with contractions near the low-latitude and low-altitude limits of their current range, and expansions to higher latitudes if migration is possible. Suitable habitat for coldwater fish in the continental United States may decline by as much as 50%, and range contractions of coldwater species could eliminate some of the world's most valuable fisheries.

Conversely, however, climate warming and changes in thermal stratification may have positive effects on cool- and warmwater species, reducing winter kills, lengthening the growth season, and increasing available habitat, both locally and regionally, if poleward migration is possible. If lakes are not nutrient-limited, productivity is likely to increase (due to increased primary productivity and growth rates); overall fish catch may increase, but there are likely to be changes in the relative abundance of fish species.

Finally, although most studies examine the effects of climate change on only a few species of fish, negative effects on one species can have an impact on the entire community. For example, a summer kill of planktivorous herring in a Wisconsin (USA) lake reduced pre-dation on zooplankton by 50%, which led to an increase in large zooplankton and intensified zooplankton grazing, causing a substantial reduction in phytoplankton abundance. In addition, while many studies concerning lake water level focus on the effects of declining levels, rising water levels due to regional or seasonal increases in precipitation could also have negative impacts. In Lake Baikal, for example, the construction of a dam along with increased precipitation led to a 1.5 m rise in the level of the lake, and a subsequent decline in fish biodiversity and production.

Effects of Climate Change on Rivers

The effects of climate change on rivers are likely to vary widely depending on latitude. Temperate rivers, like temperate lakes, will be affected primarily by temperature changes, while changes in precipitation timing and quantity could have dramatic effects on tropical rivers.

Physical Effects on Rivers

Increases in air temperature will strongly influence water temperature in many rivers (particularly smaller rivers and streams) because the surface to volume ratio of rivers is high. Increasing air temperatures will result in warmer water throughout rivers, from the headwaters to the mouth, as well as reduced oxygen levels. Rivers that are fed primarily by groundwater will be buffered against increasing seasonal variability in temperature (as groundwater temperature generally equals average annual air temperature), and may serve as thermal refugia in some areas, supplying relatively cooler water during hot seasons and warmer water during cold seasons.

Temperate rivers experience seasonal thermal cycles similar to temperate lakes, with uniform cold temperatures in winter (sometimes accompanied by ice cover) and longitudinally stratified temperatures in summer, with lower temperatures at groundwater-fed headwaters and higher temperatures downstream. High latitude rivers are already experiencing shorter periods of ice cover and earlier ice break-up, and many of the beneficial functions of ice jams (river scouring, changes in river channel morphology, flooding of riverine wetlands) may be compromised.

Flow regime is a critical component of river ecosystems. Mean flow may increase or decrease depending on changes in average precipitation, evaporation, soil moisture, and groundwater recharge, but seasonal shifts in flow may be more significant to freshwater ecosystems. Many rivers will experience altered timing or duration of high and low flows due to changes in seasonal variability of precipitation, frequency of extreme precipitation events, and timing of snowmelt. Spring snowmelts are likely to occur earlier due to warming, and winter flows are likely to increase in areas where winter precipitation falls as rain instead of snow. A shift in peak flows from spring/summer to winter will reduce the cooling effect of snowmelt on summer river temperatures.

Where precipitation increases, stream flows may increase in volume and floods may become more frequent. Extreme flooding events and landslides could remove important woody debris from rivers and destabilize river channels. Where precipitation decreases, stream flow volume may also decrease, and reductions in runoff will lower the concentrations of DOC and organic matter in rivers. Increased evaporation could also lead to reduced streamflow, even in the absence of precipitation changes. Summer and ephemeral streams in arid regions (which provide critical habitat for many animals) are more vulnerable to drying up. A reduction in natural flooding events could eliminate many of the beneficial physical effects of seasonal flooding, such as creating floodplain habitat, displacing exotic plants, and determining river channel form.

Biological Effects on Rivers

In tropical rivers, variations in air and water temperature are generally small, and water temperature is mainly regulated by shade and rainfall, rather than by cool groundwater refugia. The rainy and dry seasons of the tropics lead to large, predictable seasonal variations in precipitation and annual flooding of adjacent grasslands and forests, which provide abundant food and breeding grounds for fish. Thus, the life histories of tropical river fish are more strongly affected by changes in water level than by changes in temperature. As the rainy season draws to a close and flood-plains dry up, members of the "whitefish" guild, who are sensitive to reduced oxygen levels, retreat to the main river channel with the receding floodwaters. "Blackfishes", who are more tolerant of or adapted to low oxygen levels, remain in marginal floodplain habitats that become disconnected from the river and may even dry up completely. Some of these species, such as the lungfishes, are able to aestivate (cocoon in the mud) and breathe air when their water supply evaporates.

Climate change may affect the both the timing and extent of flooding in tropical rivers, although these effects are currently difficult to predict and will vary regionally. According to some estimates, the tropics will experience the smallest change in temperature, but the largest changes in precipitation, with rainfall becoming more variable, both within and between years. Changes in floodplain dynamics will directly affect fish populations and fisheries yield, as growth rates and overall fish catch is correlated with the area of flooded land.

Fish communities in temperate rivers will experience effects similar to those in temperate lakes. Coldwater fish that are restricted to cool refugia at headwaters during the summer will experience increased competition, reduced growth, and possible range shifts. Warmer water and decreased oxygen content at headwaters may have negative impacts on eggs and larvae often placed there. Diadromous stocks that migrate long distances during the peak of summer may experience higher rates of pre-spawning mortality because of increased metabolic needs and disease outbreaks. Even in stocks that do not perform summer migrations (such as the Adams River sockeye salmon) climate change is likely to result in a net population decline due to reduced juvenile emergence, growth and survival. Some invertebrates in northern rivers require a prolonged period of exposure to nearly 0°C water, followed by spring warming, in order for eggs to hatch. The release of warmer water in the winter from dams has resulted in massive local extinctions of invertebrates for tens of kilometers downstream, and overall river warming would be expected to have a similar effect.

Effects of Climate Change on Wetlands

Physical Effects on Wetlands

Increased air temperatures are likely to have a drying effect on many wetlands, unless increased precipitation compensates for evaporation. Shallow and ephemeral habitats, such as depressional wetlands (with no channelized flow in or out) or wetlands in arid areas could be lost entirely, especially if precipitation declines and groundwater is extracted for human needs. Overall, a drier climate is likely to lead to contractions and loss of wetland habitat, as well as increased habitat fragmentation. Conversely, increased precipitation could lead to flooding, expansion and deepening of wetland habitat, and increased connectivity. However, increased precipitation or ex-treme flooding may also lead to an increased input of sediment and pollutants, and could destroy some wetlands if vegetation or other important habitat features are completely submerged.

Arctic and subarctic bogs located over permafrost could suffer dramatic changes in hy-drological regime if rising temperatures lead to permafrost melting and wetland drainage. Increased decomposition in thawed northern peat bogs, as well as the increased risk of catastrophic fires due to drier peat, could release large amounts of carbon dioxide into the atmosphere, contributing to further global warming

Coastal freshwater wetlands are particularly sensitive to extreme high tides resulting from an increase in storm frequency or magnitude; these high tides can carry salts inland to salt-intolerant vegetation and soils, and could lead to the displacement of freshwater flora and fauna by salt-tolerant species. Rising sea levels could destroy coastal freshwater wetland communities as saline water invades, especially if these communities cannot shift inland due to development or dikes. Salt water inundation of coastal freshwater wetlands is expected to cause significant loss of wetland habitat in Australia and elsewhere.

Biological Effects on Wetlands

Ephemeral, depressional wetlands, especially those in arid areas, often harbor rare species that would be lost if these areas dry up. For example, several endemic species of fairy shrimp in California (USA) that are already severely threatened by habitat loss could disappear if reduced precipitation and increased evaporation eliminates their shallow, vernal pool habitats.

Small, temporary wetlands are the most numerous types of wetlands in many landscapes, and are often used by more species than permanent ponds The drying and loss of wetlands would reduce not only the number and size of available ponds, but also increase inter-pond distance, lowering the chances of amphibian re-colonization, since adult frogs are generally only capable of traveling 200-300 m. Drying and loss of wetlands would also reduce habitat connectivity on a regional scale, endangering migrating birds that depend upon a network of wetlands along their migration route.

Wetlands in areas with increased precipitation might suffer fewer negative effects, and may even benefit from increased wetland area and connectivity. However, some rare species that are adapted to drier, ephemeral wetlands may not be able to compete with invading species adapted to wetter habitats, and wading birds that require shallow water to feed may experience reduced access to feeding areas. Wetter, more permanent wetlands would support more fish, which prey on vulnerable tadpoles and invertebrates that usually inhabit seasonal wetlands with less predation pressure.

Managing Freshwater Reserves to Withstand Climate Change

The likelihood that individual species or communities will be able to persist in the face of global climate change depends to a large degree on how resistant (able to withstand change) and resilient (able to recover from change) they are. Because there is still a great deal of uncertainty associated with climate change predictions (especially predictions of precipitation changes), and interactions between physical and biological features of freshwater ecosystems can be complex and non-intuitive, focusing on increasing system resistance and resilience is a far better approach than trying to plan for a specific set of predicted changes.

Many common considerations in designing and managing ecosystem reserves, such as preserving biodiversity and minimizing outside stresses, will also help increase the resistance and resilience of communities to climate change. Additional considerations that are unique to freshwater ecosystems will become increasingly important in buffering systems against growing climatic and water extraction pressures.

Preserve Habitat Heterogeneity and Biodiversity

Both species and habitat diversity increase resistance and resilience to climate change, as diversity provides a greater range of stress tolerances and adaptive options. Diverse communities that have redundant species within functional groups should be more resistant to climate change because there are likely to be differences in environmental sensitivity among members within each group; functional group richness also appears to increase resistance to environmental change. High biodiversity areas may also become important as sources for re-colonizing damaged sites or colonizing new ones as the effects of climate change become more severe.

In aquatic systems, high biodiversity is often found in older or isolated habitats, in sinkholes, caves or underground habitats, and in areas with high habitat heterogeneity— especially dynamic habitats with seasonal changes in water level (i.e. river floodplains of seasonal wetlands). Many of these areas also harbor rare species, such as endemic species that have evolved in and remain restricted to a particular habitat (i.e. communities of endemic cichlid fish in African rift lakes), relict species that were restricted to isolated habitats after previous range contractions (i.e. cold stenother-mic fish that were isolated to high latitude lakes after the last ice age), or species that are highly adapted to unusual environments (i.e. cave-dwelling fish and invertebrates). In protecting some of these high biodiversity sites, rare or vulnerable species may also be protected (often a primary consideration in reserve design). Protecting rare species, especially those that are charismatic, can assist in drawing public attention and funding to conservation efforts, but strategies aimed solely at protecting one species may detract from the more desirable goals of protecting ecosystem function and increasing resistance and resilience to climate change.

Areas where natural physical barriers separate biota (i.e. impassable waterfalls), and transition zones between different habitats or ecosystems may also harbor high biodi-versity. Protecting transitional zones has the added benefit of accommodating possible range shifts due to climate change, and can help preserve diverse habitat types. Protecting a variety of potential habitats may help increase resistance and resilience in vulnerable species; for example, protecting an array of natural ponds with a wide range of sizes and hydroperiods will help ensure that amphibians have access to suitable breeding sites regardless of climatic variation. If possible, replicate sites of a particular habitat type should be protected to safeguard against the complete loss of critical species or communities if one site is damaged beyond repair by an extreme climatic event.

Although diverse communities may be more resistant and resilient to climate change, it is important that high biodiversity not be used as the sole criterion in selecting sites for conservation. An equally important goal is protecting communities that perform valuable "ecosystem services", such as relatively low-diversity wetlands that provide flood protection, water filtration and other services that are likely to become increasingly important as climate variability and extreme events increase. Protecting sites with lower biodiversity also helps maintain functioning ecosystems over broader regions of the globe, and preserves distinct evolutionary lineages that can provide fuel for future evolutionary innovation. Preserving sites harboring high biodiversity can still be a valid conservation goal in many areas, and diversity indices can be extremely useful in choosing between different sites *within* a habitat type (rather than using biodiversity indices as an absolute guide in choosing *between* different habitats). However, considering ecosystem services and other potentially beneficial features of low diversity sites is an important aspect of planning for climate change.

Protect Physical Features Rather than Individual Species

Aquatic ecosystems differ from many other ecosystems in that they are usually governed by "bottom-up" rather than "top-down" dynamics—in other words, much of ecosystem function is determined by basic physical features such as water flow, channel morphology, and nutrient balance, rather than by species assemblages. Protecting flow patterns, water quality, and water quantity will go a long way towards protecting biodiversity in freshwater habitats, whereas conservation efforts that focus solely on preserving particular species or groups of species without considering wider physical features of the system may be doomed to failure. In many cases, the function of a species in a freshwater ecosystem is actually more important than its identity; for example, plants are essential components of some aquatic habitats (i.e. floodplain vegetation and aquatic plants in shallow lakes), but the exact species of plant may be less important than the physical features it provides.

The physical features of rivers, lakes and wetlands are expected to undergo a number of changes as a result of climate warming and precipitation variability. Removing barriers to water flow, maintaining healthy, forested river basins, and reducing the input of nutrients and toxic substances will increase the likelihood that freshwater ecosystems will be able to adjust to climate change. For example, removing levees and other barriers to the lateral expansion of rivers could prevent the loss of critical edge habitats and the species that depend on them, by allowing new floodplains to be established if average river flows increase or extreme precipitation events become common.

Preserve Habitat Connectivity

Connectivity is an important feature of many freshwater ecosystems, as it can help preserve flow regimes, promote ecological integrity, and allow migrating animals to move between different habitats at various life history stages. Connectivity is important not only between different freshwater habitats (i.e. between rivers, lakes, and wetlands), but also along the length of rivers, and between freshwater habitats and subterranean systems or groundwater sources. Maintaining connectivity will become even more important in some areas as the effects of climate change increase, because connectivity may provide animals with access to thermal refugia, or allow them to migrate to more suitable habitats.

Although some species may be able to adapt to climate change in their current habitats, warmer waters will force other species to move into cool, thermal refugia, where temperatures remain below their thermal tolerance limits and metabolic demands are lower. Many species (i.e. coldwater fish) already rely on thermal refugia at certain times of the year, and these species are likely to become even more dependent on these refuges for year-round survival. Headwaters of rivers and any areas where temperature-buffered groundwater enters a system should be protected, and vegetation over bodies of water should be maintained to provide cooling. Maintaining or increasing connectivity between cool refugia and the rest of an ecosystem should be considered, as this may help provide additional species with access to these areas.

In many cases, where thermal refugia do not exist or other climate-related changes make an animal's environment uninhabitable, the only option may be migration to more suitable habitats. Because migration of aquatic animals is already severely limited by the direction of connections between water bodies, preserving or improving the possible migration corridors that do exist may be important in some cases. Rivers and other aquatic ecosystems that allow movement in the north-south direction, as well as freshwater systems spanning altitudinal gradients are particularly valuable migration corridors. In addition, areas near the current range limit of species should be protected (in the direction of expected migration). Consideration should be given to protecting areas that are currently not of interest, but that species are likely to migrate into, or marginal habitats (i.e. areas that are too cold for most species) that are likely to be improved by climate warming.

Although maintaining connectivity between habitats and ecosystems may have a number of beneficial effects in terms of allowing native species to adapt to climate change, the increasing threat of invasion by exotic species makes connectivity in aquatic systems a more complicated issue. Because invasion by exotic species can potentially have devastating effects on ecosystems, the risks of invasion should be weighed on a site-by-site basis against the vulnerability of native populations to climate change, and the necessity and feasibility of migration to other habitats. In cases where species are particularly sensitive to temperature changes and are likely to experience local extinctions due to thermal stress, and where migration corridors leading to more suitable habitats exist, the benefits of connectivity may outweigh the risks of invasion. In other areas, however, the risks of invasion may be so severe that allowing one sensitive species to be lost would be preferable to endangering the entire community.

Protect Sites from Human Pressures and Exotic Species

Protecting reserves from outside stresses (particularly stresses that tend to reduce diversity) will become increasingly important as local climates become more variable, because stressed systems display reduced resistance and resilience to change. Human stresses, such as overexploitation and poor land use practices, should be reduced as much as possible. In choosing between a group of candidate sites for new reserves, it may sometimes be advisable to avoid sites that are already severely degraded or likely to be subject to intense human pressure, although restoration of some ecosystem functions is possible.

It is also critical to increase efforts to prevent access of invading species to reserves, and to eliminate or control harmful non-native species already present. Many systems are likely to become more vulnerable to invasions, as thermal barriers that previously excluded invaders will be removed, and communities that are already stressed by climate change are invaded by warmer-adapted species. Even in sites that are well-protected from all other stresses, non-native species can wreak havoc on biological communities; for example, native fish populations in the pristine Blindekloof River in South Africa have been nearly devastated by invading largemouth bass.

Unfortunately, in some systems, preventing access of motile, invasive species may conflict with the goal of maintaining connectivity to allow seasonal or climate-induced migrations. Isolating vulnerable habitats from other freshwater ecosystems may be feasible in some cases, but building barriers that disrupt flow to prevent access of exotic species may do more harm than good. In cases where the risk of invasion is relatively low and migration is important to resident species, maintaining current levels of connectivity while enacting careful monitoring of ecosystems (to allow early, aggressive management responses to invaders) may be a suitable approach.

Manage Entire Watersheds and Regulate Extractive Water Use

Freshwater ecosystems are intricately connected to their drainage basins, and downstream or lowland rivers, lakes and wetlands can be extremely sensitive to distant, upstream disturbances. Deforestation, agriculture, and other pressures on terrestrial communities that drain into water bodies can alter the quality, quantity, and temperature of water in all freshwater systems downstream of the disturbance.

Because freshwater ecosystems are so intricately connected to one another and to the terrestrial systems that surround them, common reserve considerations such as reserve size and buffer zones are less applicable. For example, Kruger National Park in South Africa is a very large reserve that protects significant downstream portions of the rivers that flow though it; however, populations of several species of fish that are protected within the park have declined as a result of upstream activities outside of the reserve. Although reserve size and buffer zones may be applicable in limited, small-scale cases (i.e. no-take fishing areas and buffer zones to protect seasonal breeding grounds in Lake Tanganyika), protected habitat patches will generally provide only short-term solutions. Freshwater reserves will not be secure unless upstream threats are removed by managing (or in rare cases protecting) the entire drainage basin.

Managing entire watersheds rather than simply protecting aquatic elements or habitat patches will become increasingly important as the effects of climate change intensify and are magnified by interactions with human stresses. Human population growth is likely to lead to increased deforestation, agriculture, industrial development, and urbanization within watersheds. Models of land use and climate change in South Africa predict that abrupt, future changes in local land use will have a far greater impact on freshwater hydrology than gradual effects of climate change. Importantly, the stresses caused by these habitat alterations often exacerbate the effects of climate change. For example, deforestation near freshwater ecosystems eliminates cooling, vegetative cover over streams and reduces the input of large, woody debris. The loss of shading further increases water temperatures that are already rising due to global warming, and the lack of large, woody debris eliminate in-stream refugia. In addition, the loss of forest cover may lead to increased terrestrial and aquatic evaporation, and reduced soil moisture and water infiltration. During heavy precipitation events, streams already experiencing unusually high flows would receive increased runoff due to low soil infiltration, as well as increased inputs of sediments and pollutants.

Perhaps even more important than projected increases in habitat alteration are the dramatic projected increases in human water extraction. Population growth will lead to increased water extraction for agricultural irrigation, direct consumption, and industrial development. Increasing human water extraction is likely to be one of the primary stresses on freshwater ecosystems in coming years, and is expected to greatly outweigh global warming in affecting water supply through at least 2025. Increasing human water demands on freshwater ecosystems that are particularly sensitive to climate change are likely to lead to more frequent water shortages and conflicts over water use. For example, Egypt relies on the Nile river basin for over 95% of its national water budget. Based on one scenario of global climate change, models predict that a modest decline in precipitation (20%) and rise in air temperature (4°C) would nearly halt flow of the Nile (98% flow reduction), and human demands would be likely to outweigh ecosystem needs in determining the fate of the remaining water. Clearly, management schemes aimed at protecting freshwater ecosys- tems from climate change must take human water needs into account, and attempt to manage the timing and magnitude of water extractions.

Restore Degraded Sites

Because freshwater ecosystems are strongly influenced by activities in the surrounding landscape and increasing amounts of water are being extracted for human needs, most drainage basins have already experienced some degree of degradation. Restoration of degraded sites holds great promise for freshwater ecosystems, in terms of both improving the ecological integrity of damaged systems (thus increasing resistance and resilience to future change), and providing tools to smooth the recovery of ecosystems from future damage caused by climate change. Successful restoration techniques include neutralizing acidified lakes and rivers by applying lime, and restoring the hydrology of wetlands by removing impediments to flow. Exotic species such as predatory game fish have been successfully removed from some freshwater systems, but in many cases, controlling aquatic community composition has proven difficult. All of the ecosystem problems mentioned above could potentially be caused or exacerbated by climate change (i.e. lake acidification, changes in hydrology, invasions of exotic species), and lessons learned from current restoration projects can help guide future responses to climate change. However, it is important to keep in mind that restoration efforts are unlikely to be successful after certain thresholds of damage have been crossed, such as the collapse of fisheries or permanent cultural eutrophication of lakes.

Rather than attempting to restore systems after they have already been degraded by climate change, it may be possible to perform active interventions to ameliorate the effects of climate change or to directly protect vulnerable species. Adding to water tables below drying wetlands or draining excess water from flooded habitats has been proposed, and inter-basin transfers of water have been suggested as a possible adaptation to climate change in Africa and South America. However, interbasin transfers result in the intermixing of diverse faunal communities that were previously isolated, and may have unforeseen effects on native communities. In addition, tropical rivers and lakes (for which many of these projects have been proposed) are particularly vulnerable to the negative effects of interbasin transfers. Flow conditions below dams could possibly be manipulated to relieve some of the negative effects of climate change, for example by releasing bursts of water to simulate flooding events or releasing water from multiple reservoir layers to control temperature. However, although these actions may help alleviate some of the negative effects of climate change on systems already impacted by dams, erecting dams solely to control water flows would cause far more damage to the ecosystem than the original alteration in flow regime caused by climate change.

Some of the more controversial suggestions put forward for helping species and communities adapt to climate change involve transplantation of animals. Although protecting severely threatened species *ex situ* (i.e. in zoos or aquaria) until a suitable habitat for reintroduction becomes available is a sound idea, using artificially-aided dispersal to move aquatic creatures over watershed boundaries to cooler waters could lead to the extinction of existing fauna, as the transplanted species would be pre-adapted to warming waters and could have a range of unanticipated effects on the native community. Similarly, the often-devastating effects that non-native fisheries stocking can have on native fauna would seem to make introducing better-adapted non-native species to replace waning stocks a bad idea. Transplanting vulnerable species to new, created habitats where no native community exists (i.e. uninhabited, enclosed basins) would eliminate the risk of damaging existing biotic communities, but creating functioning, self-sustaining ecosystems from scratch is likely to prove difficult. Re-introduction of species to their native habitat may be a useful approach in cases where an extreme weather event destroys existing populations.

Use Adaptive Management Strategies to Maintain Flexibility

Because many aspects of climate change are unpredictable and climatic variability is likely to increase, it is critical to maintain flexible conservation goals and strategies. Rather than striving to maintain current distributions of species and preserve current ecosystems, managers should allow for and even assist in the adaptation of species and communities. Static management practices with fixed policies designed to protect particular species should be changed to adaptive management strategies aimed at protecting ecosystem processes.

Adaptive management is based upon the recognition that uncertainty is inherent in all natural processes, and the expectation that management practices will change over time. Passive adaptive management involves adjusting management practices based on what is learned from the results of past practices, but learning about the underlying system is not an explicit goal. In contrast, active adaptive management is somewhat like performing a (hopefully) well-designed experiment; it involves forming multiple hypotheses about how the system will respond (to climate change and/or management practices), choosing strategies to systematically test and learn about the underlying hypotheses, monitoring systems closely to evaluate responses, and adapting future management practices based on what was learned about the system. Management practices in which a single (often arbitrary) strategy is chosen teach us little about the underlying system, and it is often impossible to know exactly what aspects of the strategy led to its ultimate failure or success.

Obviously, there are some cases where experimenting with the response of a system is not an option; some strategies may be irreversible, or may be too risky if populations are already endangered. However, adopting active adaptive management practices now, while the effects of global climate change are only beginning to be felt in many areas, may provide enough time to learn about the underlying processes governing how a particular system responds to change, and provide an understanding of how to best manage the system in the face of global climate change.

Selecting and Implementing Adaptation Strategies

It illustrates some of the conflicts that may arise in choosing a management strategy to sustain freshwater ecosystems in the face of global climate change. For example, should the focus be on protecting high-diversity sites that are more likely to be resistant to climate change, or on protecting low-diversity sites that perform ecosystem services that will become more valuable as climate change continues? Should connectivity between habitats be maintained or even increased to allow migration, or should habitats be isolated to prevent invasion by exotic species? Applying a general list of conservation goals blindly to a freshwater ecosystem ignores the specific strengths and weaknesses of the system, and does not account for the relative importance of various threats. Performing a careful, site-specific analysis that takes all known factors into account and evaluates how to best meet conflicting goals is likely to be the best approach to designing a successful adaptation strategy.

Characterize and Monitor Species/Systems

One of the most important steps in selecting an adaptation strategy for climate change is to characterize which life history stages, species, communities, or physical features in your location are most vulnerable to changes in average climate, changes in climate variability, or extreme events. This can help determine the scale of management that is necessary. For example, if only one species of coldwater fish in a series of lakes is declining and the rest of the community appears to be resistant to climate change, a small-scale conservation strategy such as protecting thermal refugia and maintaining potential migration corridors to pristine lakes at higher altitude may be sufficient. However, if increased precipitation and runoff from deforested, agricultural lands is causing a dramatic increase in sediment load and nutrient content, along with large-scale changes in species composition, distribution and abundance, a more ambitious basin-wide management plan may be called for.

Attempting to predict which species will be most vulnerable to climate change before large effects are observed can be very useful, as initiating careful monitoring before effects are noticeable will provide a baseline against which future changes can be compared. While predicting exactly how species will respond to change can be difficult, numerical methods that combine assessments of climatic sensitivity (thermal tolerance, etc.) with general vulnerability (life history traits, current knowledge of the species, taxonomic uniqueness, overexploitation) may help highlight species for further consideration. Identifying and protecting keystone species is also important in order to prevent cascading effects that would alter the entire community.

Performing careful and continuous monitoring of both biological and physical indicators of change is critical, as reactions to climate change can be delayed in many plants and animals. For example, directly monitoring the water table and hydrology of temperate wetlands may be critical in preventing habitat loss, because by the time the vegetation shows a response to environmental change it may already be too late to save the system.

Develop Strategies for Protecting Freshwater Ecosystems

A serious potential threat, and one unique to aquatic ecosystems, is the anticipated rise in human water needs, mainly due to population growth and increasing development. Pressures caused by water extraction and climate change will almost certainly interact, exacerbating the effects of climate change on ecosystems and possibly increasing human water needs further due to increased temperatures and evaporation. Proposed adaptation strategies for human water resource management include "demand-side" adaptations, such as price incentives for conserving water, enforceable water efficiency standards, and increased irrigation efficiency, as well as less environmentally-friendly "supply-side" adaptations, such as building more dams. The Intergovernmental Panel on Climate Change recommends using Integrated Water Resource Management (IWRM) to adapt to increasing water resource demands. In this process, all stakeholders are included in potential considerations of supply- and demand-side actions before a decision is made, and the situation is continuously monitored and re-evaluated. Unfortunately, IWRM does not consider maintaining aquatic ecosystem function as one of the goals of water management (i.e. aquatic ecosystems are not considered a "stakeholder"); environmental damage is only included as a potential negative side effect of some actions.

A vast improvement over this water management strategy in terms of maintaining ecosystem health is ecologically sustainable water management, which strives to protect the ecological integrity of freshwater ecosystems while meeting current and future human needs for water. This strategy has been applied primarily in cases where one or more large dams already exist on a river and there are conflicts between increasing human water extraction and ecosystem flow needs, a problem that is likely to become more common in the future due to climate change. Methods of addressing water conflicts commonly include altering patterns of surface and groundwater extraction, increasing efforts to improve water efficiency, and changing temporal patterns of water release from dams. All stakeholders are involved in the process of estimating ecosystem flow requirements (seasonal base flows, high and low flows, rates of rise and fall), determining current and future human water needs, identifying incompatibilities (i.e. in seasonal or regional needs), and collaboratively searching for solutions. A critical component of this management strategy in terms of adapting to climate change is the inclusion of short-term management experiments (i.e. injecting treated wastewater back into groundwater reserves) to test the effectiveness of various management options as climate change continues to alter water flows, ecosystem needs, and human water demands. The method also calls for the development of an on-going adaptive management strategy to continuously monitor and respond to ecosystem changes.

This method has proven effective in many cases, such as in the management of the Green River Dam in Kentucky, USA. This dam was managed to provide recreational opportunities in the reservoir during the summer and water storage capacity to protect against flooding during the winter. As a result, large quantities of water were released from the dam during the fall, which biologists believed were disrupting downstream prey aggregations and dispersing mussel larvae during the fall breeding season. After examining ecosystem and human needs and identifying the incompatibilities (mainly related to the large outflow of water in the fall), the management of the dam was altered to release a steady, low volume of water throughout the fall, along with big bursts in November coinciding with natural storm events. The reservoir level was still lowered before the winter, when large storms could cause flooding, and the new schedule provided a more natural flow regime for downstream wildlife. This case provides an especially “tidy” example of ecologically sustainable water management (particularly because human water extraction was not a major issue), but the method has also proven effective in far more complicated cases where multiple governing bodies were involved and the goals of various stakeholders differed widely.

Perform Integrated River Basin Management

Although a primary goal of ecologically sustainable water management is maintaining the ecological integrity of freshwater systems, the main focus is on managing the water itself. As described previously, freshwater ecosystems depend on more than just the water they contain—rivers, lakes and wetlands are intricately connected to all of the terrestrial systems within the drainage basin surrounding them. To protect freshwater ecosystems that are showing large-scale changes in response to climate change, or systems that are likely to be vulnerable (due to land alteration or other anthropogenic stresses), the entire drainage basin must be managed.

One-sided decisions to adopt fixed river basin management plans are likely to fail if unexpected climatic changes occur. For example, the tri-nation development authority in charge of the Senegal River basin in the Sahel region of West Africa initiated a plan several decades ago to convert large areas of the river basin to irrigated rice production, in an effort to reduce the countries' dependence on foreign imports. Two dams were constructed to provide irrigation water, but during the 1960's, the Sahel began to experience a severe and long-term drought that current climate models suggest may be permanent, constraining the amount of water available for flooded rice paddy agriculture. Over the last several decades, the basin has suffered increasing desertification, due in part to climate change and in part to the abandonment of cleared rice paddies. As a result of water shortages and river basin degradation, large numbers of local residents have migrated to urban areas. Proposed changes in river basin use include switching from industrial rice production back to local, village-based agriculture utilizing cereal and grain crops with low irrigation needs, as well as a more ambitious plan to convert areas of the basin to agro-forestry, which involves mixed vegetable production within plots of reforested land; this reforestation would help lower air and water temperatures, and halt desertification. The reforestation plan also recruits unemployed urban residents in an attempt to reverse recent patterns of massive demographic shifts to urban areas.

In order to avoid the necessity of restoring river basins that have been severely degraded by climate change and failed management, the basin-wide needs of ecosystems and local communities should be considered, as well as the potential impacts of future climate change, before enacting any management plan. Integrated River Basin Management (IRBM) is a method of balancing basin-wide ecosystem needs with human water resource needs to achieve economic, social, and environmental goals. One of the major problems with basin-wide management is that many river basins cross national boundaries; worldwide, there are 261 major transboundary rivers that drain 45% of the Earth's surface, account for 80% of the planet's river flow by volume, and are home to 40% of the world's population. Transboundary rivers can make water management more difficult, but management of these systems is essential to future economic and political stability.

With IRBM, the needs and expectations of all "water stakeholders" (local community members, civil authorities, water and fishery resource managers, scientists and conservationists, and representatives of the private sector) from all countries are assessed jointly, a basin-wide authority is created, monitoring methods are developed, and an adaptive management plan is initiated. Decisions can be made locally, but must be in accordance with basin-wide strategy. Key aspects in determining the success or failure of IRBM are actively promoting public involvement (through appropriately-scaled local discussions suited to the target audience) and ensuring sustainable funding for the basin-wide authority (often through water taxes, reductions in water subsidies, or international funding programs). Because it employs adaptive management strategies, IRBM is able to continuously respond to the effects of climate change. In addition, because management is basin-wide, IRBM provides the ability to protect climatic refugia and areas that are particularly vulnerable to climate change, as well as to minimize damaging land use practices and other human stresses that are known to interact with and exacerbate the effects of climate change.

Although system-wide conservation can present difficulties in terms of reaching consensus, and can be costly and slower to achieve noticeable results than local campaigns to preserve endangered species, system-wide approaches ultimately conserve the func-tional value of the river basin and provide a sustainable basis for future conservation. Successful IRBM in Costa Rica, in which fossil fuel taxes and payments from a private hydroelectric plant compensate upstream landowners for the maintenance and restoration of forest cover, has improved water quality and quantity for downstream towns, farms, and industries, and reduced sediment accumulation at the hydroelectric plant. In South Africa, IRBM led to the creation of the Working for Water Programme, a massive project that protects biodiversity while creating jobs for 18,000 people clearing invasive, water-hungry plants from several river basins.

Within a managed river basin, priority should be given to protecting upstream sub-basins and headwaters; these areas serve as thermal refugia, providing cool, oxygen-rich groundwater to many species not found downstream, and are often the destination of migratory fish and other animals. In addition, upstream areas affect the flow regime and water quality of all areas further downstream, and are often less degraded and easier to protect. It is also important to protect side channels and backwaters, which serve as refugia and spawning grounds for a range of animals, and provide corridors to important floodplains. Finally, other downstream areas should also be protected, as these areas are necessary for the migration of diadromous species, possess abundant water resources, and are often more productive and species-rich than upstream areas. However, downstream areas are also more frequently subjected to human pressures such as water extraction and land use, so protecting these areas may involve restoration or creative approaches to resource management.

Preserving the Ecological and Evolutionary Driving Forces

Freshwater ecoregion conservation (ERC) involves protecting relatively large units of water and surrounding land that contain distinct assemblages of natural communities sharing many of the same conditions. These ecoregions are defined by the similarity of the communities and conditions contained within them, and can include one or more drainage basins. Many freshwater ecoregions encompass a number of basins and cover vast tracts of land, such as the Amazon River and Flooded Forests ecoregion and the Yangtze Rivers and Lakes ecoregion.

The main difference between freshwater ecoregion conservation and IRBM is that ecore-gion conservation focuses primarily on preserving biodiversity, while IRBM emphasizes balancing environmental, economic, and social needs. In addition, ecoregion conservation is often applied to areas of high-biodiversity importance, in order to preserve these exceptional areas while addressing and analyzing the ecological and evolutionary driving forces of biodiversity. Because of its focus on protecting high-diversity communities and ecosystems (which are likely to be more resistant and resilient), and because the large scale of ecoregion conservation allows managers to control most non-climatic stresses affecting freshwater ecosystems, ERC is particularly well-suited to buffer systems against the effects of climate change. Additional benefits of an ecoregion approach include the ability to address the conservation needs of wide-ranging species or species that require particularly large areas of habitat, and the ability to address threats that operate across an entire ecoregion with a single, coherent approach.

In relation to climate change, freshwater ecoregion conservation may provide the most comprehensive and holistic approach to providing adaptation options (thermal refugia, migration routes, etc.) and minimizing anthropogenic stresses, thereby allowing regions of paramount importance in biodiversity to resist, recover from, and/or adapt to global climate change. However, because many river basins throughout the world are heavily populated, a more balanced approach (such as IRBM) that considers socioeconomic factors in adaptive management of river basins may be more likely to provide long-term protection; as human water demands increase and human populations experience more direct effects of climate change and extreme climatic events, interest in protecting ecosystems is likely to wane and management schemes with the sole goal of protecting freshwater ecosystems and biodiversity may lose support. Ideally, IRBM could be applied to more heavily populated or degraded basins within a freshwater ecoregion, but overall management of the region could be guided by ecological principles aimed at increasing resistance and resilience to climate change, thus combining the benefits of the two approaches. Ultimately, however, any management strategy for dealing with global climate change will simply buy time until either rapid, anthropogenic climate change ceases or species are no longer able to adapt and massive extinctions result. Management strategies can only provide long-term protection of freshwater ecosystems if the root causes of climate change are addressed and solutions are enacted on a global scale.

References

Anderson, D.M., Overpeck, J.T., and Gupta, A.K. (2002). Increase in the Asian southwest monsoon during the past four centuries. *Science 297*:596-599.

Appleberg, M. (1998). Restructuring of fish assemblages in Swedish Lakes following amelioration of acid stress through liming. *Restoration Ecology 6*:343-352.

Atkinson, D. (1995). Effects of temperature on the size of aquatic ectotherms: exceptions to the general rule. *Journal of Thermal Biology 20*(1/2):61-74.

Avila, A., Neal, C., and Terradas, J. (1996). Climate change implications for streamflow and streamwater chemistry in a Mediterranean catchment. *Journal of Hydrology 177*:99-116.

Bahls, P. (1992). The status of fish populations and management of high mountain lakes in the western United States. *Northwest Scienc*e *66*:183-193.

Brinson, M.M., and Malvarez, A.I. (2002). Temperate freshwater wetlands: types, status, and threats. *Environmental Conservation* 29(2):115-133.

Westmacott, J.R., and Burn, D.H. (1997). Climate change effects on the hydrologic regime within the Churchill-Nelson River Basin. J*ournal of Hydrolog*y 202:263-279.

Williamson, C.E. (1995). What role does UV-B play in freshwater ecosystems? *Limnology and Oceanograp*hy 40(2):386-392.

7

Ecosystem Approach to Freshwater Ecosystem Management

In the past, the use and management of the natural riverine environments has largely been of a sectoral character, which has not only resulted in an inefficient use of freshwater ecosystems, but also in a widespread transformation of riverine landscapes and the degradation of water and ecological quality and fluvial dynamics. However, the parallel, yet not interconnected, development of two integrated and ecosystem-oriented management approaches show the potential to constitute a new management and water policy framework to guide water management in the direction of improved and more-integrated management practices in the years to come. These management approaches are the Ecosystem Approach (EsA) of the Convention on Biological Diversity (CBD) and the Water Framework Directive (WFD) of the European Union (EU). The general objective of this chapter is to analyse the consideration of the Ecosystem Approach in the Water Framework Directive with an emphasis on the correspondences (and the potential discrepancies) on the conceptual level of both approaches. .

Ecosystem Approach

The Ecosystem Approach (EsA) of the Convention on Biological Diversity (CBD) is a strategy for management of land, water, and living resources that promotes conservation and sustainable use in an equitable way. The Conference of Parties (COP) has adopted the EsA as a primary framework for action under the CBD. The COP 5 adopted the 12 principles, including annotations to the rationale and five operational guidelines, in May 2000 in Nairobi. In 2002, the COP 6 (The Hague) assessed the development of further refinement of its principles. The seventh Conference of Parties in Kuala Lumpur in 2004 agreed on the implementation guidelines to each principle for refinement and elaboration of the EsA. The COP noted that the principles were not always precisely worded. Therefore, advice and elaboration to overcome the problems of clarity and interpretation were added. However, the EsA still has to be regarded as a general framework for holistic decision-making and action. The EsA includes ecosystem processes, functions, and interactions among organisms and their environment. It recognises that humans and cultural diversity are an integral component of ecosystems. Further, the ecosystem approach of the CBD involves a wide range of stakeholders at different scales of application.

The 12 principles are the key elements of the ecosystem approach. The principles are perceived as being complementary to each other in an interactive context. The principles imply that objectives are a matter of societal choice (Principle 1), and recommend that management should be decentralized to the lowest appropriate level (Principle 2). Effects of activities on neighbouring ecosystems are to be considered (Principle 3). Ecosystems should be understood and managed in an economic context (Principle 4). However, conservation of ecosystem structures and functions is a priority target (Principle 5). Management needs to be carried out within limits of ecosystem functioning (Principle 6). Management must include consideration of the appropriate spatial and temporal scales (Principle 7), while objectives have to be set for the long-term (Principle 8). Furthermore, management must recognise that some changes are inevitable (Principle 9). The above objectives are the preconditions for an appropriate balance between conservation and use of biodiversity (Principle 10). To achieve these objectives it is necessary to consider all forms of relevant information (Principle 11) and involve all relevant sectors of society and scientific disciplines (Principle 12). In this section, we will use the 12 principles to assess similarities and divergences between the EsA and the WFD.

Principles of the Ecosystem Approach of the CBD

The 12 Principles of the Ecosystem Approach of the CBD are listed below:

1. The objectives of management of land, water and living resources are a matter of societal choice.
2. Management should be decentralized to the lowest appropriate level.
3. Ecosystem managers should consider the effects (actual or potential) of their activities on adjacent and other ecosystems.
4. Recognizing potential gains from management, there is usually a need to understand and manage the ecosystem in an economic context. Any such ecosystem-management programme should:
   a) Reduce those market distortions that adversely affect biological diversity;
   b) Align incentives to promote biodiversity conservation and sustainable use;
   c) Internalise costs and benefits in the given ecosystem to the extent feasible.
5. Conservation of ecosystem structure and functioning, in order to maintain ecosystem services, should be a priority target of the ecosystem approach.
6. Ecosystems must be managed within the limits of their functioning.
7. The ecosystem approach should be undertaken at the appropriate spatial and temporal scales.
8. Recognizing the varying temporal scales and lag-effects that characterize ecosystem processes, objectives for ecosystem management should be set for the long term.
9. Management must recognize that change is inevitable.
10. The ecosystem approach should seek the appropriate balance between, and integration of, conservation and use of biological diversity.
11. The ecosystem approach should consider all forms of relevant information, including scientific and indigenous and local knowledge, innovation and practices.
12. The ecosystem approach should involve all relevant sectors of society and scientific disciplines.

Clearly, there is no single way to achieve the EsA for management of land, water, and living resources. It depends on local, regional, national, and global conditions. In order to address management issues in different social contexts, the principles have to be translated in a flexible manner. The EsA is a holistic concept including an integrated land use planning that seeks the appropriate balance between nature conservation and use of biodiversity. This implies a "high degree of complexity in management", which includes ecological, socioeconomic, cultural, and political issues. The term ecosystem is generally used more as a construct of a complex system rather than a geographic entity. This term is not to be used in any particular spatial unit or scale and it should be determined by the frame of reference.

In the context of freshwater management and the preservation of aquatic and water-dependent ecosystems, several coordination mechanisms with other international conventions and bodies were established, and the CBD adopted a specific programme of work (programme of work on the biological diversity of inland water ecosystems). In this policy area, the close cooperation with the Ramsar Convention appears particularly relevant. Both conventions have launched, inter alia, a so-called River Basin Initiative with the objective to stimulate the international diffusion of innovative management approaches to freshwater ecosystem management. CBD's current programme of work on inland waters calls for the application of the EsA and underlines the necessity to reach integrated forms of management.

Ecosystem Approach to the Management of European Rivers

The area of the enlarged European Union is a share of a relatively small continent, with mostly a temperate, humid climate and a long coastline. Although the EU is densely populated, it has a high proportion of agricultural areas. The rivers of the EU are numerous, yet relatively small and short (EEA 1994). The larger rivers are located in the central part of Europe reaching from the Vistula in the East, to the Rhone in the West. The larger rivers are mostly transboundary rivers with a high share of settlements along their banks. The countries with a long coastline in relation to size of area, e.g., Norway, the UK, Italy, and Greece, have large numbers of small rivers and concentrations of settlements along their coastlines.

Ecological Conditions of of European Rivers

The river flows have an annual pattern following a seasonal pattern of precipitation and are regionally influenced by the thawing of snow and ice, resulting in a regionally typical flow regime, e.g., the Alpine flow regime or temperate lowland regime in Southern Europe. With extensive swamps, lakes, and forests, the natural fluctuation is attenuated naturally and by man-made structures. The rivers are intensively used for hydropower generation in the mountainous regions and for inland shipping and hy-dropower along the major rivers in the central lowlands from the Volga to the Rhone. As a result of the heavy physical modifications, inland fishing has become of minor importance compared to the end of the 19th century.

Water abstraction in the EU is, on average, relatively small in relation to total renewable resources, but there is considerable regional variation. Freshwater availability is low (i.e. below 5000 m^3/cap) not only in Mediterranean countries as in Spain or Greece which one would expect, but also in Western European countries, e.g., Belgium, the UK, and Germany, as well as in the Czech Republic and Poland. Freshwater use is dominated by public water supply and industrial use. Agricultural use is important only in the Mediterranean countries, varying between 50 and 80%, whereas in the rest of Europe agriculture consumes less then 10% of water abstracted. In the most highly stressed countries, there are signs of groundwater overuse with groundwater tables declining and saltwater intrusion on coastal aquifers. Water abstractions continue to be high in the Mediterranean countries, while they declined slightly in Western Central Europe and considerably in the Eastern accession countries to the EU.

Groundwater pollution from nitrates is a major problem in Western Central Europe and pesticides in a broad range of countries. The nutrient and organic pollution of inland and coastal waters improved with heavy investment in wastewater treatment plants since the 1970s. Actual reductions occurred in Central and Eastern Europe, but there are still problems with non-point sources, particularly from agriculture. As a consequence, nutrient discharges into the seas changed with improvement of emissions from point sources into the North Sea and the Baltic, but little improvement in the Mediterranean and the Black Sea occurred. With respect to hazardous substances (heavy metals, pesticides, and other organic pollutants), there has been improvement in the Nordic and Western European countries with respect to mercury and cadmium. The sale of pesticides has been declining over the last ten years, but the level of sales is still high in Western Europe, particularly the Netherlands.

Besides water pollution, European natural riverine environments in general – and wetlands in particular – came under heavy pressure. Today, almost all the larger rivers are to some degree regulated, and many rivers were even massively transformed in the context of industrialisation and infrastructure development in the past decades. The effects and the intensity of pressures depend on the type of wetlands (marshland, floodplains, etc.) and the human interventions, as well as on the population density in the respective areas. The long history of European settlement, permanent agriculture, and industrialization has changed the natural freshwater ecosystems of regions including the species of flora and fauna, which are an integral part of them. Because of the drainage of lowland areas for agriculture and urban development, wetlands were particularly threatened. In addition, the massive transformations of river systems for navigation, flood control, power generation, and water storage led to an impoverishment of river ecosystems, in particular as regards floodplain ecosystems. Some figures might serve as an illustration: In Greece, 60,000 ha of wetlands disappeared in the late 1940s and 1950s due to the construction of dams and drainage tunnels and a further 390,000 ha because of partial drainage. In France, the sites of 78 major wetlands were degraded more than 85% in the period from 1960 to 1990. Bulgaria had 200,000ha of wetlands at the beginning of the century and only 11,000ha have survived until today. However, many important wetlands were preserved and political attention to this topic has been increasing in the last years, even though many of the remaining wetlands are fragmented and much altered compared to the original conditions (Jones 1996). There are several river restoration projects in many European countries with the aim to regain hydrological dynamics and to remove constructions such as dams, channels, embankments, etc. However, the effects remain somewhat isolated and significant improvements have proven to be difficult to achieve, mainly due to economic, institutional, and political reasons.

European Water Framework Directive

The competences of the European Union, as well as the duties and the rights of the different European institutions, are laid down in the treaties of the EU. The founding treaties from the 1950s have been substantially reformed on several occasions. The latest amendment is the treaty of Nice that entered into force in 2003. Relevant implications of the different amendments for environmental policy-making explicitly empowered the EU to move into the environmental policy area with the Single European Act (1987), the increasing application of (qualified) majority decisions in the European Council (again enlarged with the treaty of Nice), the strengthening of the role of the European Parliament, and the emphasis on the subsidiarity principle. The subsidiarity principle underlines that member states should be given freedom to enforce EU environmental policy regulations through means that they prefer. While the EU clearly has gained importance in environmental policy-making in the last decades – in fact, the impact of the EU on a member state's domestic environmental policy is more than significant – there still are some particularities that are worth keeping in mind.

First, there are (legal and factual) limits of the decision-making power of the European Union relating to environmental policy in general and water management in particular. For instance, the EU is not in a position to precisely prescribe several implementation tools with the consequence that instruments such as water-related taxes, water charges, or land-use regulations can barely be regulated or harmonized throughout Europe. Binding decisions would still demand unanimity in the European Council. Consequently, the WFD (like all other EU directives) does not only reflect the state of expertise on water management and the respective national experiences, but also the imponderables stemming from complex European decision-making processes. In this context, it is also worth noting that the member states partly pursue very different regulatory approaches to environmental management in general and to water management in particular. For instance, there is a long-lasting and sharp debate on the issue of whether environmental legislation should regulate the sources of potential damages or focus on environmental quality. These contrasting approaches lead to different regulatory preferences and the EU environmental policy has suffered for a long time from somewhat inconsistent regulatory approaches. Even though this controversy has somewhat softened in the last years, most of the EU directives – and the WFD is no exception – are still characterized by these different positions.

Second, all EU directives have to be transposed into national law and, furthermore, need a national or regional implementation on the ground. The transposition into national legal systems, by whatever national arrangement, is often a time-consuming and complex process. In this context, it is particularly important that the regulation style has somewhat changed in the last years with the effect that, meanwhile, the EU environmental policy relies more on framework directives with the objective to better coordinate the 'regulation patchwork' in the different environmental policy fields (water, air, etc.) and to leave room for detailed regulations on the national level. In practice, however, the quality of national approaches to implementation varies a great deal between countries and from case to case. This is particularly true for the EU water policy regulations because non-compliance with a couple of water-related directives or delayed implementations were significant in the past. The consequences of these tendencies for our discussions are twofold. On the one hand, EU directives should generally not be equated with national law because the latter is generally more comprehensive and precise than EU directives, in particular regarding administrative procedures and the allocation of competences. On the other hand, one of the main striking features of EU environmental and nature protection policy is the widespread lack of implementation with the greatest shortfalls occurring in cases of those EU directives which demand huge investments, innovative administrative procedures, and/or new interactions of political and administrative actors. In face of the challenging character of many requirements of the WFD, the 'real' implications of the directive are not easy to predict, but implementation obstacles are to be expected.

Water Policy of the European Union and the Development of the WFD

In the past three decades, the water policy of the European Community had a core consisting mostly of regulations against urban and industrial pollution to protect human health and to harmonize national policies for competition reasons. These regulations consisted of two types of directives:

a. Directives regulating the ambient quality of waters according to the intended uses were mostly is sued in the early phase of the Community's environmental policy: Drinking Water Abstraction Directive (75/440/EEC), Freshwater Fish Directive (78/659/EEC), Shellfish Waters (79/923/EEC), Groundwater Directive (80/68/EEC), Bathing Water (76/160/EEC), and Drinking Water (80/778/EEC).

b. An emissions based approach under the Dangerous Substances Directive (76/464/EEC) and a number of daughter directives for specific pollutants/ industries. The Dangerous Substances Directive was conceptualised as a framework directive with two lists of substances: List I for substances for which the EU must come up with emission limit values and List II for which the member states must develop their own reduction programmes. The EU managed to agree on 8 daughter directives for 18 substances.

Yet, the general approach remained a matter of conflict among member states. A group of member states, e.g., the United Kingdom, favoured the ambient quality approach where the consensus has to be achieved on the European Community level and the resulting reduction obligation has to be developed on an individual or member country level. This was opposed by countries, e.g., Germany, committed to a uniform emission limit value approach based on the precautionary principle and the best available technologies. The conflict was resolved when a combined approach could be agreed upon. The role of emissions limits uniformly applied gained momentum when two major water pollution control directives were passed in 1991. The Urban Waste Water Treatment Directive which sets limits for municipal treatment plants and the Nitrates Directive (91/676/EEC) which obliges the member states to introduce programmes of reduction for vulnerable zones that they have identified.

The coverage of the water directives has been patchy and their implementation record varies considerably. The Drinking and Bathing Water directives are considered successful because they gained public attention in the member states. The Urban Waste Water Directive is considered effective as well despite the high costs of 150 bil. Euros estimated for 1993 to 2005 which it has imposed on urban residents, while the Nitrate Directive with a more complex structure of instruments and implementation is regarded as being poorly effective.

The debate about sustainability in the early nineties and the limited success of the EU water policy, including the results of the Dobris Assessment on the European Environment in 1992, gave way to review of the whole approach of water management in the mid-nineties. It started with a proposal of the Commission in 1993 to introduce a directive about the ecological quality of surface waters, but this proposal was not pursued further. Instead, the Commission presented a communication to Council and Parliament in 1996 that contained an analysis of the existing directives and developed the basic structure of a water framework directive. It already had the main elements of the WFD: broader goals, the integration of all water types, the integration of water quality and quantity, the organization on a river basin basis, transparency, and public participation. In addition, it included practical improvements: consolidation of the existing quality-oriented directives, and less reporting, yet more effective monitoring. In 1997, the Commission presented its proposal for the WFD as an input into the European Community decision-making procedure with the Council members trying to avoid additional specific environmental standards and the Parliament trying to make the directive stricter. It took four years to negotiate the final text of the framework directive, which differs from the original Commission proposal by granting more time and more lenient exceptions to the member states in implementing the directive, and by leaving the administrative set-up to the member states instead of having river basin organisations as a standard and obligatory solution. The Community obligation to provide a risk assessment for hazardous substances with the aim of a progressive reduction was introduced, but for the WFD no agreement on those substances to be included could be reached.

Objectives, and Instruments of the Water Framework Directive

As the name indicates, thse Water Framework Directive of the European Union issued in October 2000 is a directive which establishes a framework for Community action in the field of water policy, in particular for the management and use of freshwater and coastal water ecosystems and for those terrestrial ecosystems directly dependent on aquatic ecosystems in the member countries within the European Community. The previous water related directives of the European Union, which were specific with respect to pollutants, protected water use. In setting either ambient water quality or source specific emission standards, the WFD defines the objectives for the protection and sustainable use of all the water ecosystems in a holistic manner. It sets a timetable to achieve the goals, and provides a framework for the implementation in terms of organization and instruments. The overall objective is to achieve good ecological and chemical status of surface water and good quantitative and chemical status of groundwater. For those waters currently not achieving the good status, a river basin plan and a program of measures have to be developed to achieve this objective by 2015. Furthermore, the deterioration of those water bodies currently having good or better status has to be prevented.

The Protection Objective

The novelty of the protection objective is that it is based on ecological criteria, i.e. the requirements of fauna and flora concerning their aquatic habitat, and no longer on certain human uses, e. g., for bathing or drinking water, as in previous directives. The directive requires good status for surface water in terms of ecological and chemical status and regulates the process of defining these criteria which are referred to as a very good ecological quality. The categories of water bodies (rivers, lakes, estuaries, and coastal) are classified into types according to their geological, geographical, and hydrological characteristics and then described in their natural state, i.e. without anthropogenic degradation. As a high portion of the European water bodies does not reach that high ecological quality, the objective of the WFD is then to achieve good ecological quality (the second in a five category classification system) within 15 years after the WFD entered into force, i.e. in 2015. The ecological status is evaluated by biological and hydro-morphological criteria. The management of this process is required to take place in the member countries on a river basin basis.

For chemical status, a three-tiered procedure is incorporated:

(a) Priority hazardous substances (Art. 16 (2)) have been identified by the EU in 2001 (first list of 33 substances) and they have been added as Appendix X to the WFD, for which zero emissions are targeted for 2020 to protect ocean waters;

(b) For a number of priority substances (Art. 16 (7)), which are discharged in significant amounts, the Commission will identify a list of priority substances and propose ambient quality objectives, based on the protection of aquatic communities and human health; and

(c) For the remaining substances, member states set ambient quality standards according to a procedure set.

If the assessment of the ecological and chemical status of a water body leads to the conclusion that it does not achieve good status, the member states have time until 2009 to start a set of activities and combine them to an action programme in order to achieve good status by 2015.

In Art 4 (3) an exception from the above formulated EU-wide objectives has been made for the category of artificial water bodies (channels, surface mining lakes, as defined in Art 2 (8)) and heavily modified water bodies (as defined in Art 2 (9)). While there is a consensus that there are no natural reference conditions for artificial water bodies, the implications of the exceptions for the heavily modified water bodies (HMWB) from the overall objective of the WFD are seen with suspicion. The designation of HMWB is possible if the achievement of good status would require hydro-morphological changes which would have significant adverse effects on human water uses such as navigation, power generation, storage for irrigation, flood protection, and land drainage (Art 4 (3a)), or the beneficial effects of the modification cannot be achieved by other means technologically feasible, at reasonable costs, or with better environmental effects (Art 4 (3b)). All these conditions have to be met and the associated processes of screening and assessing have to be standardized in a guidance docu-ment. If they are all met, an alternative set of objectives has to be defined in the form of a Maximum Ecological Potential based on a comparison with the closest surface water body and in consideration of all feasible mitigation measures. The Maximum Ecological Potential is then used to derive the Good Ecological Potential, which will be used as the objective to achieve for these water bodies by 2015. An exception relates mostly to physical alterations of rivers such as dams, locks, dredging, and bank fixation. The mitigation measures consist of low cost measures which do not require changing the basic alteration e.g., dam removal, but do require changes in water quality discharges.

For groundwater, the WFD establishes a ban on further deterioration and aims to ensure good status as well. "Good Status" is defined in quantitative terms as achieving a balance between abstraction and recharge and in terms of groundwater quality. For defining the quantity dimension also the status of terrestrial ecosystems which depend directly on the groundwater body has to be considered. The quality dimension is defined by the chemical status of the groundwater, by introducing ambient quality objectives, and by obliging the member states to reverse trends in groundwater pollution. There are only a few EU-wide ambient standards in force (nitrates, pesticides) in separate directives; consequently, there is a lack of standards for the remaining pollutants and there is no regulation in the WFD for lack of a consensus. The WFD in Art 17 (2) requires the Commission to develop a proposal for a separate directive to deal with the remaining ambient standards and the criteria to define the starting point for trend reversals in groundwater pollution. The Commission published a proposal in 2003 (COM (2003) 550 final) which will leave the process of identifying the pollutants and of defining the specific standards up to the member states for the bodies of groundwater at risk.

Sustainable Use

The use of water and of water bodies is covered in a variety of ways without explicitly referring to sus-tainability. However, sustainability is strongly implied.

The WFD is most definitive about the sustainable use of groundwater, as sustainable use requires the balance between abstractions and recharges in order to attain good status for groundwater (Art 4 (1bii)). This objective is measured against variations in the groundwater level and the direction of water flow to detect the inflow of saltwater.

There are no explicit objectives concerning abstractions from surface water bodies, but the requirements for the protection of freshwater habitats imply that minimum water flows are protected and total abstractions remain below that level. A similar argument can be made for those discharges into freshwater that will still be allowed after the protection objectives have been followed (nutrients, heat, etc.) and are not subject to zero emissions goals.

An indirect reference to the sustainability of use is made by the requirements of cost recovery for water services (Art 9 (1)), including the environmental and resource-based costs. In the economic aspect, the main focus is on the recovery of financial costs for infrastructure that provides drinking and irrigation water and sewage collection and treatment, because there are a number of regions in Europe where the recovery of these costs from the user deviates from the polluter (user) pays principle. A higher recovery is expected to lead to a corresponding adjustment of water consumption and waste-15 water discharge. This can be considered as the application of conventional well-established economic reasoning. More innovative is the intention of the WFD to apply this principle to a wider range of water and river uses, called water services in the directive. In Art 2 (38) they include not only water abstraction, treatment, and discharge, but also storage and impoundment as well. The advising "Working Group Economics" has specified this as including hydropower and use of water bodies for shipping. The cost recovery principle is established in a relatively soft manner. The WFD requires that until 2010 the member states oblige the main water using sectors (agriculture, industry, and public water supply) to contribute their adequate share to the costs of water services which is less than a full cost recovery, taking into account the social, ecological, and economic implications of the principle.

Common Implementation Strategy

The WFD as a framework directive leaves considerable space for decisions to be made by the member states. The process of organising the water management on a river basin basis is basically left to the member states.

Only the milestones of the planning process are specified, but here the basic principles of integration, of public participation, the application of the polluter-pays principle, the objective of good status, and the use of economic analysis are laid down. The member states have to transform the directive into national law, designate the River Basin Districts (RBDs), and ensure that they meet the deadlines. To ensure a comparable level of implementation they have to adhere to the Common Implementation Strategy of the Commission and cooperate with each other in the international RBDs.

Because the WFD leaves considerable autonomy to the member states with respect to a number of topics in implementation which are technical in nature, the Commission and the member states (plus Norway) decided to devise a common strategy for the implementation of the WFD. The strategy consists of the creation of an expert network, the common sharing of information, as well as the development of guidance documents and their testing in pilot river basins.

After these guidance documents were completed and accepted by the water directors of the member states, these documents and the efforts of the member states to use them were put into a testing exchange. One important area relates to the definition of good ecological status. In this regard, developing a comparable classification of water bodies and of their classification with respect to ecological status resulted in the establishment of a network of sites to assess the comparability of the national classification systems and the performance of a meta-analysis of the results on a community wide basis. A similar effort goes into the exchange about and testing of the methodologies for integrated river basin management.

Implementation by the Member States

As the WFD went into force in October 2000, the then 15 member states had to transpose it into national legislation by the end of 2003 (Art 24). The new acceding states had to achieve this goal by May 2004, their date of membership. Not all member states have achieved this goal. An EEB survey in May 2004 found that in 10 of the 25 current member states, the WFD has been transposed into national law. In September 2004, the European Commission established its WFD scoreboard, which showed that 15 member states had fulfilled their obligation to the satisfaction of the Commission, and five countries had not submitted information.

At the same time, the member countries had to identify the river basins within their territories and assign them to individual River Basin Districts (RBD) (Art 3 (1)) and to identify appropriate competent authorities (Art 3 (2)).

The WFD establishes the river basin as the regional basis for the definition of the objectives and as a basis for the action plans to achieve the objectives. As this basis does not exist in all member countries (only a few countries have river basin management agencies), the WFD has to establish a process for defining water bodies and river basins in order to form River Basin Districts. While, the water bodies are geographic areas that are defined hydrologically (rivers, lakes, aquifers, and artificial, transitional and coastal waters) and for reporting purposes (differences in ecological quality), the RBD is consideration of the 12 Principles of the Ecosystem Approach in the European the legal management unit for a set of water bodies. A river basin includes not only rivers and lakes but also aquifers, and artificial, transitional, and coastal waters. A River Basin District consists of these river basins (one or more), which are not only made up of the hydrological system, but also of the land and sea (Art 2 (15)). In member states with long coast lines, usually relatively short rivers exist and more than one river basin will belong to a RBD (e.g., Sweden, Italy, and Greece), whereas other countries with short coastlines and sizable country areas will usually have one River Basin District for one basin. This is explicitly formulated as guidance in Art 3 (1).

The member states designate administrative units, which are responsible for the functions assigned to the RBD in the WFD, and ensure appropriate coordination for the organizations (national or subna-tional) contributing to the adequate fulfilment of the RBD functions. The RBDs have responsibilities for the aquifers in river basins and for the coastal waters as well. Groundwaters have to be identified and assigned to the nearest and most appropriated RBD, an explicit classification of the delimitations is not required (Art 3 (1). The coastal waters are defined as all waters landward from a line 1 nm seaward of the baseline (Art 2 (7)).

Because of the geography of the EU and its member states, a number of river basins are international and in these cases, the WFD requires the member states to establish an international RBD (Art 3 (3)). In a case in which the international river basin is shared with non-member states, the EU member states concerned "shall endeavour to establish appropriate coordination with the relevant non-Member States". The aim is the achievement of the objectives of the WFD throughout the RBD (Art 3 (5)). Thus, the WFD covers the area of the EU member countries including the countries that acceded in May 2004. It has an influence on river basin management of transboundary rivers, which have a share of their basin inside and outside the EU territory, as is the case for the Danube and the Baltic and Finnish International RBDs. The member states have to provide a list of their competent authorities and of the competent authorities of international bodies in which they participate by June 2004 (Art 3 (8)). The WFD Scorecard of the Commission shows that by September 2004, 12 member countries had submitted the required information.

Functions of the River Basin Districts

The RBD will perform a number of functions to implement the requirements of the WFD on the river basin level for which the directive has a set of deadlines. The member states are required to ensure that the designated authorities will perform these functions. In Germany, the existing water management Laender ministries and agencies responsible for water management have been assigned specific tasks by the Laender. Public law agreements about cooperation among the Laender have been signed by the Laender to aggregate the results of every step from the Laender level up to the RBD level. The functions basically conform to the standard progress of steps in any planning process.

With the relatively tight schedule, the WFD creates enormous pressure within the water management agencies to assemble all the necessary data, to make them useful for the objectives, and prepare them for further analysis, planning, monitoring, and reporting.

For every single step, a guidance document has been developed The water bodies have to be characterised according to eco-region types including reference conditions according to chemical and hydro-morphological characteristics. It may include a provisional identification of heavily modified water bodies. The current and foreseen anthropogenic pressures have to be identified and their impacts on environmental quality have to be assessed. This implies that their influence has to be predicted and the risk of the water bodies failing to meet their environmental status objectives assessed. If the answer to the risk assessment is uncertain, then in 2005 to 2006 a monitoring phase is required.

The register of protected areas under Art 6 does not provide the power to designate protected areas, but it requires the RBD to summarise areas already protected under separate directives. There are five categories of protected areas, four of which are protected under water specific directives (Nitrates Directive, Bathing Water Directive, Shellfish Waters Directive, and Surface Water Abstraction Directive).

The last component of the analysis of the status quo is the economic analysis of water uses, which is characterised as the first milestone for the economic analysis. Here, the importance of water and water use for the economy is to be assessed, the development of the economy as a basis for future pressures is to be analysed, and the current level of cost recovery is to be estimated as well.

In some cases, the member states may also need to establish programmes of investigative monitoring, e.g., where the reason for any exceedances of the environmental objectives are unknown or where a risk of failing the environmental objectives is indicated but an operational monitoring has not already been established. Combined with the setting up of the environmental objective, for which there is no explicit deadline, the monitoring serves as the basis for the next step of developing the Programme of Measures.

The core instruments of the WFD are the Programme of Measures (PoM) according to Art 11 and the River Basin Management Plans (RBM plans) according to Art 13. The RBM Plan as the final output is conceptualised as a summary of all the planning activities: description of the status, significant pressures and impact of human activities on the status, a list of the objectives, a summary of the programme of measures, a register of detailed programmes and management plans for sub-basins, and a summary of the public information and consultation activities. Because this plan has to be reported to the Commission in five years, the description of its content is still rather general. As these directives have already been in force for quite a period, the PoM of the WFD thus serves as a mechanism to enforce their implementation. The reviews of the European Commission on the implementation of the two major directives showed considerable deficits. The designation of vulnerable zones by the member states was mostly more restrictive than assessed by the Commission. The installation of treatment plants – with considerable variation -remained behind the directive's obligations. Similar results were reported for the action programmes of the Nitrates directive.

The next set of measures concerns the control over abstractions and discharges. The WFD asks for a register of abstraction of freshwater, artificial recharges of groundwater, and point pollution discharges as well as their prior authorisation and a periodical review. Exemptions are allowed based on risk evaluations.

In addition, the PoM should include those measures, which ensure the recovery of costs for water services including environmental and resource costs as required by Art 9. Costs of water services include the treatment and delivery of freshwater, and the collection, treatment, and disposal of waste-water. This applies to use by private household, commercial, and agricultural users. Implied – not explicitly stated in the WFD- is that the prices should be linked to the water quantity used and/or the pollution produced to provide an incentive structure to the user. In a note to the European Parliament, the Commission defined the various cost categories: financial, environmental, and resource costs, but the WFD remains rather vague concerning the degree of cost recovery and whether differences for the three categories are allowed. The guidance document on economic analysis is helpful in providing an overview of the use of economics and in the assessment of the current levels of cost recovery to be reported by the end of 2004.

If all of these measures combined are not sufficient to achieve the good quality as required, then additional supplementary measures have to be developed, nationally or on a river basin level. Major shortfalls are expected in meeting the goals for chemical status of surface and groundwater because of pollution from non-point sources (mostly from agriculture). The existing programmes of member states that deal with the nitrates directive (e.g., to support changes in agricultural practices, cropping pattern, vegetation cover, and type of agricultural land use) will have to be modified and expanded. The trends towards converting wetlands will have to be stopped and wetlands will have to be restored. To deal with deficiencies in ecological status, a number of water bodies will have to be changed towards a more natural state, by recreating wetlands, natural morphological conditions, by building more fish ladders, by removing barriers and a number of similar measures as well. The list of these measures will depend on the economic tests of designating HMWB, the cost effectiveness of the whole programme, and on the derogation test of Art 4 (4) and (5) based on the assessment of the disproportionateness of the associated costs. The final programme of measures will depend on a number of calculations (economic tests) to be taken between 2005 and 2009, for which the results are far from being predictable.

The management process is rather complex in those member countries that have to reorganise their water management administration. These member countries do not have an administrative structure with river basins as the major organising principle. The need for cooperation among federal units is exacerbated in federally structured countries (e.g., Germany). Another change required in most member countries is the need to involve nature protection agencies via the requirement to protect wetlands and to include protected areas with nature protection objectives in the analysis of the current situation.

In addition, the WFD requires that the decision-making process in the implementation phase is characterised by active involvement of all interested parties (Art 14 (1)). It focuses on the river basin management plan and thus on the River Basin District, and not on the objectives and instruments of the WFD itself. Article 14 distinguishes three levels of participation in terms of the binding character: 'information', 'consultation', and 'active involvement'. Member states have to ensure consultation and access to background information, but only have to encourage active involvement. In addition, the WFD distinguishes between 'interested parties' and the wider 'public' in the sense that information and consultation are prescribed for both categories, whereas active involvement only for interested parties. The obligation is most specific for the information level. The member states have to complete the following in three phases: First, a timetable and work programme for the production of the RBM plan, including a statement of the consultation measures to be taken, have to be published and made available for comments to the public, including users. Then, there is an interim overview of the significant water management issues identified in the river basin that also has to be published. Finally, the member states are required to publish and make available for comments the draft copies of the RBM plan, at least one year before the beginning of the period to which the RBM plan refers. Access shall be given to background information for the draft, upon request. For the next level of participation and consultation, the WFD requires that the documents listed above are made available for comments and that the authorities shall allow six months to submit comments in writing. Other forms of consultation, e.g., hearing and public fora, are not required, but they are not excluded. The organisation of these levels of participation is up to the member countries.

On the third level, the WFD obliges the member countries to encourage the active involvement of all interested parties, but does not elaborate this point further in the article. The guidance document on public participation defines as the core of active involvement that "interested parties participate actively in the planning process by discussing issues and contributing to their solution". These activities are likely to play a role in the early phases of the implementation process. Further forms of participation – shared decision-making and self-determination- are not required by the WFD, but are classified as best practice in the guidance document. It refers to a number of examples in member countries which can be considered as best practice.

The River Basin Management plans have to be decided upon and published in 2009. The PoM has to be established in the same year and made operational three years later in 2012 (Art 11 (7)). By 2010, the member states ensure that the pricing policies are in place to contribute to the environmental objectives of the WFD. They decide whether they are applicable for all or individual river basins. The relevant planned measures will be reported within the PoM (Art 11 (3b)) in 2009. There is a period of three to five years, during which the measures are expected to work based on the appropriate administrative arrangements in each RBD to achieve good ecological status for the water bodies. Then, all of these activities have to be reviewed in 2015 and repeated every six years thereafter (Art 5 (2), 11 (7), and Art 13 (7)).

Ecosystem Approach in the Water Framework Directive

In the following section, the consideration of each of the twelve principles of the Ecosystem Approach in the WFD will be discussed in detail. The discussion is largely based on the text of the WFD; in addition we consider several guidance documents mentioned above. Furthermore and to the extent possible, a rough assessment of the national implementation prerequisites are given in order to derive tendencies concerning the expectable effects of the directive on the ground.

While both concepts relate to the governance and management of natural resources, the potential application area of the EsA is much broader because it relates to all kinds of ecosystems. It is a worldwide concept under the umbrella of a global convention, and it is intended to develop basic management principles for all management levels (from the local to the global level). Consequently, the wording of the EsA – in spite of some recent clarifications –has to be rather unspecific and is not meant to lay out concrete objectives to be achieved by the parties. Therefore, we encounter the problem that the EsA allows a number of different interpretations and the twelve principles are not accompanied by a set of assessment or 'consideration' criteria. Consequently, a precise assessment of the extent of consideration of a specific principle does not appear feasible. Therefore, our 'consideration rankings' must remain on the level of general evaluations and the identification of tendencies.

On the other hand, the WFD shows several particularities stemming from the fact that the directive is a legal piece of the European Union and should not be equated with national law. For our discussion, it is particularly important to note that the implementation of the directive is still in a very early stage and, more generally, the effectiveness of all EU directives arises from the interaction of the European law with national regulations and institutions. Insofar, our discussion clearly shows a preliminary character because the effects of the directive on the implementation level are not yet fully predictable.

Principle 1

The objectives of management of land, water and living resources are a matter of societal choice *Rationale:* Different sectors of society view ecosystems in terms of their own economic, cultural and society needs. Indigenous peoples and other local communities living on the land are important stakeholders and their rights and interests should be recognized. Both cultural and biological diversity are central components of the ecosystem approach, and management should take this into account. Societal choices should be expressed as clearly as possible. Ecosystems should be managed for their intrinsic values and for the tangible or intangible benefits for humans, in a fair and equitable way.

As a general rule, water administration in most EU member states traditionally holds considerable influence regarding both the targets and the way water management is carried out, whereas the immediate influence of society on management decisions is comparatively low. However, there is a huge variety of institutional approaches to water management within the borders of the European Union: several member states already show a systematic and formal participation of stakeholders via water parliaments on the level of river basins (e.g., France) whereas in other countries (e.g., Germany and the Netherlands) the direct involvement of stakeholders is rather the exception than the norm and, if at all, more common on the local level and/or restricted to selected management tasks. Against this background, it is important to consider that the WFD explicitly calls for public participation in river basin management and the directive will thus potentially demand new procedural approaches to the integration of the public and stakeholders in the decision making process, at least in those member states whose water laws and practice do not – or only very limitedly – provide for similar regulations.

The phrase "public participation" does not appear explicitly in the WFD. However, it requires public participation as an integral element of the river basin planning and management process. Preamble WFD para. 14 acknowledges that the implementation success of the directive relies, inter alia, "...on information, consultation and involvement of the public, including users". The specific norm addressing the role of "societal choice" in the water planning process is Art 14 of the WFD, which specifies the requirements relating to the involvement of the general public and the relevant stakeholders. However, it is worth mentioning that the public participation requirements of the WFD should not be interpreted as an isolated norm, but rather in the context of relevant international conventions and complementary European directives.

In this context, the Aarhus Convention is the most relevant piece of international law because Art 14 of the WFD is – albeit not explicitly mentioned in the directive – the legislative implementation of this convention in respect to the interface between water management and environment protection in EU Consideration of the 12 Principles of the Ecosystem Approach in the European WFD legislation. Aside from the Aarhus Convention, the European Directive on Strategic Environmental Assessment (SEA Directive, 2001/42/EC) complements the WFD as regards public participation in the planning procedure. In particular, a strong interaction of the WFD and the SEA Directive on the implementation level is expectable because the Programme of Measures will also fall under the regime of the SEA Directive. The Programme of Measures is to be implemented by the member states according to Article 11 WFD in order to achieve the environmental objectives of the directive.

In Article 14 WFD 'Information' is understood to mean access to background information that is used for the development of the draft river basin management plans (required by Art 13 of the WFD). It is an open question whether the WFD's provision for 'information' should already be classified as 'participation' because information can be provided by means of press releases, brochures, etc. Yet, these means are a one-way communication flow because people do not have the opportunity to express their opinion.

Article 14 of the WFD calls for 'consultation' in relation to the three key steps of the planning process. In the context of the WFD, consultation aims at "learning from comments, perceptions, experiences, and ideas of stakeholders". Article 14 WFD explicitly mentions written consultation as a minimum standard for implementation, but member states could – and probably should – consider oral or active consultation in addition.

In contrast to mere information and consultation, active involvement implies that stakeholders are invited to contribute actively to the planning process by discussing issues and contributing to the development of solutions. Clearly, active involvement is much more in line with the spirit of Principle 1 of the EsA than a participation strategy strictly limited to 'information' or 'consultation' only. According to Art 14 (1), member states "shall encourage the active involvement of all interested parties in the implementation of this Directive, in particular in the production, review and updating of the river basin management plans." Therefore, the encouragement of active involvement is explicitly contemplated for, but not limited to, the whole planning cycle regarding the river basin management plans. The notion "shall encourage" can be judged as a rather weak and unspecified wording. Although it leaves much room for different interpretations and styles of implementation in the member states, active involvement is not a mere voluntary exercise. Member states have to report their respective activities. Furthermore, some commentators expressed the opinion that the need to gain support from various actors for water protection measures required by the WFD will make some forms of active involvement in the planning process unavoidable regardless of the weak wording of Art 14.

There are no legally binding documents that specify the requirements of Art 14 of the WFD, but the present guidance document on public participation provides some indications and points out broad principles of "why, what, who and how" stakeholders should be involved. In addition, several member states did (or intend to) develop their own national guidelines on the procedural requirements with the objective to facilitate and harmonise implementation. However, a few open questions remain on the implementation level: which groups are 'interested parties' (respective stakeholders in the sense of the guidance document), what are the effects of the different geographical scales (local, regional, national, and transboundary) in the design of participation procedures, how should the active involvement be shaped, which competences should eventually be transferred directly to user groups, etc. In particular, the 'problem of scale' for the organisation of participation procedures is far from being fully solved and experience in large-scale or even transboundary participation in Europe is scarce. However, context specific approaches in the member states are expectable. These approaches are largely dependent on the respective public participation traditions and administrative styles.

Generally, the implementation of the Art 14 is a highly challenging issue in those member states where the self-perception of the water administration is far from the recognition of the need to actively involve the society in water management decisions. In addition, in many cases a lack of sufficient administrative resources in terms of staff, financing, and expertise is to be expected. Therefore, an effective implementation of Art 14 is not granted per se. On the other hand, there are a couple of encouraging examples in the member states which demonstrate the feasibility and the fruitfulness of active involvement of stakeholders in river basin planning mechanisms.

Summing-up: Principle 1 of the EsA is broadly considered in the WFD in that "societal choice" in the WFD is mainly equated with public participation on the three different levels mentioned in Art 14 of the WFD. In addition "societal choice" may already exist presently through the traditional role of the water administrations and the elected representatives in the democratic European political systems. Compared to previous EU water legislation, the WFD is the first piece of water law that specifically calls for public participation and underlines its importance for the success of the directive on the ground. In contrast to other (rather technical) provisions of the directive, the wording of Art 14 remains somewhat unspecified and leaves much room for different interpretations and approaches in the member states.

Clearly, member states restricting the implementation to the minimum requirements of the directive will not experience a dramatic shift in their water management decision procedures. Furthermore, which participation procedures will stand in time, particularly in the face of the expected complex – and often controversial – negotiations within the River Basin Districts, is an open question. Some guidance can be found in the relevant document of the Common Implementation Strategy.

Principle 2

Management should be decentralized to the lowest appropriate level.

Rationale: Decentralized systems may lead to greater efficiency, effectiveness and equity. Management should involve all stakeholders and balance local interests with the wider public interest. The closer management is to the ecosystem, the greater the responsibility, ownership, accountability, participation, and use of local knowledge.

One of the most critical challenges of water management is to match the physical boundaries of the ecosystems with the boundaries of the political or administrative systems, which hold the competences in respect to ecosystem management. Non-integrated policies or management strategies address only a part of the water system, such as a river stretch or a single type of pollution source. Not considering the broader context and the interdependencies within the natural system would seriously risk of insufficiently taking physical externalities into account. This includes upstream and downstream effects relating to water quality, quantity, or adjacent land-use. Consequently, it is a broadly shared view in the international debate on water management that the river basin is the most appropriate scale for an integrated management approach. The argument that river basin management – in combination with catchments/ watershed initiatives – is the best way of protecting water resources in an integrated manner is broadly reflected in the work programme on inland waters of the Convention on Biological Diversity, too.

Because the WFD promotes the river basin as the adequate management scale, the directive generally is in line with the requirements of principle 2 of the EsA. Even the wording of the WFD is close to the CBD's ecosystem approach because the directive explicitly calls for "decisions to be taken as close as possible to the locations where water is affected or used". The concept of river basin management is institutionalised in Art 3 of the WFD. These River Basin Districts are the key spatial unit for all environmental objectives and the specification of measures under the WFD, comprising the development and implementation of river basin management plans, the programmes of measures, the river basin and economic analyses, public information and consultation, monitoring programmes, etc. River basins covering the territory of more than one member state are to be assigned to an international River Basin District. In case of international River Basin Districts within the territory of the European Union, member states are required to ensure the necessary coordination and may, for this purpose, use existing institutional structures such as interna-28 tional river commissions, transboundary working groups, etc. In the case of River Basin Districts extending outside of EU territory, the relevant member states 'shall endeavour to establish adequate coordination with the relevant non-member states with the aim of achieving the objectives' of the WFD.

Albeit the river basin is the general norm for institution building, there are exceptions in case of small river basins. According to Art 3 (1), such smaller basins may be combined with larger river basins or joined with neighbouring small basins to form larger River Basin Districts where appropriate. Herewith, the WFD acknowledges that the institutionalisation of river basin management should not only take into account the hydrological circumstances but administrative requirements also. The consequence of this regulation is that in member states with long coast lines, where usually relatively short rivers exist, more than one river basin will belong to a RBD, whereas other countries with a short coastline and sizable country areas will have one RBD for each basin.

It is obvious that the WFD puts the main emphasis on surface waters, while groundwater is not of prime importance in the process of institutionalisation. According to Art 3, groundwater bodies (and coastal waters) that do not fully follow a particular river basin shall be identified and assigned to the nearest or most appropriate River Basin District. Therefore, the wording of Art 3 with respect to groundwater bodies clearly leaves room for national implementation and it is obvious that the directive does not require particular institutions for the governance of groundwater bodies.

Generally, the WFD does not specify how member states should implement the River Basin Districts internally in terms of organizational set-up and coordination of the relevant administrations. In particular, member states are not obliged to set up a specific type of organisations – such as river basin authorities – but rather they are only obliged to ensure "the appropriate administrative arrangements, including the identification of the appropriate competent authority, for the application of the rules of this Directive within each river basin district..." (Art 3 (2)). This relatively vague requirement is the result of a highly controversial debate relating to an earlier draft of the directive that intended to make river basin authorities obligatory. Compared to this early proposal of the European commission, the current formulation of Art 3 was chosen to better reflect national administrative traditions and sub-national allocations of competences. In particular, the federal political systems within the European Union that do not possess their own river basin management tradition strongly opposed a compulsory obligation to create river basin authorities.

In this context, it is worth mentioning that there is a huge variety of the hitherto existing approaches to river management in the European Union. Although some member states possess a long river basin management tradition with well established administrative structures (e.g., the French model with water agencies accompanied by river parliaments), there are only very few experiences in a couple of other member states where river basin organisations, if existing, are limited to sub-basins. Germany, for instance, is an example of a country where water management is extensively organized around political-administrative units. Therefore, the rather vague requirement of the WFD ('appropriate administrative arrangements') apparently reflects this institutional diversity and the diverging national points of departure.

Against this background, the establishment of administrative units, which are responsible for the functions assigned by the WFD, means for some countries assigning the function to already existing river basin organisations (e.g., France, Spain, and United Kingdom). For other countries, it means devolving national authority to RBDs (e.g., Italy, Greece, and Finland) or creating coordinative mechanisms for federally structured member states with large RBDs (e.g., Germany and Austria). In a 2003 survey, Nilsson et al. found that 96 RBDs have been designated in the member states of the enlarged EU (without data on Italy, Greece, Malta, Cyprus, and Wallonia), plus those in cooperating Norway and Romania. Of these RBDs, 29 are international: 16 among the 25 members and 13 with non-members. The non-members are Norway, which signed an agreement with the EU and participates in the common implementation strategy, and the neighbouring countries in the East (Russia, Belarus, the Ukraine, and Moldavia) and in the Balkans, forming the Danube river basin.

In addition to the institutional promotion of regionalisation, the WFD also shows greater sensitivity to regional circumstances in another way. In reaction to past criticism of EU-wide uniform standards, the directive emphasizes more regional differences on the level of environmental objectives and better respects regional differences in the regional natural endowments and ecosystem qualities. For instance, the environmental objectives – to be determined according to the provisions of Art 4 of the WFD – may vary because the reference parameters for the "very good status" of surface waters may vary according to the different river types. These different reference types will result in regionally differentiated objectives and, consequently, context specific programmes of measures.

While the experiences with institutional innovations in reaction to the WFD still are limited, it is foreseeable that several member states will organize the implementation of the directive in a logic of "nested" institutions. Although the planning process is dominantly organized as a bottom-up process, the form of coordination within the River Basin Districts is not yet clarified. In Germany for the purpose of preparing for the river basin management plans, the larger River Basin Districts are sub-divided into smaller sub-basins in order to ease data gathering, the river basin analysis, and the administrative procedures associated with the development of the plans. The criteria used for the sub-division of the River Basin Districts are partly hydrological, but administrative criteria also played an important role which is justified as a way of minimizing administrative upheaval and easing coordination with other sector plans and programmes. However, it remains unclear how the (potentially con-flictive) process of aggregating sub-basin plans will be managed in the future, in particular, the determination of cost-efficient programmes of measures in cases of upstream-downstream externalities. Some critics warn that by focusing on the sub-basin in this stage of implementation, coordination across the whole river basin could potentially be neglected. On the other hand, the sub-division of large river basins, such as the Rhine, clearly shows an advantage in that (local and regional) stakeholders can be more easily involved in the planning process.

Summing-up: Principle 2 of the EsA is explicitly considered in the WFD but the directive is rather vague in terms of institutional and organisational requirements. Member states are currently in an early process of institution building; therefore the relative effectiveness of different options is not yet clear. A huge variety of institutions in the EU member states is, as expected, strongly influenced by the previous institutional approaches to water management.

Principle 3

Ecosystem managers should consider the effects (actual or potential) of their activities on adjacent and other ecosystems.

Rationale: Management interventions in ecosystems often have unknown or unpredictable effects on other ecosystems; therefore, possible impacts need careful consideration and analysis. This may require new arrangements or ways of organization for institutions involved in decision-making to make, if necessary, appropriate compromises.

Traditionally and not different from most national water policy approaches, the water legislation of the European Union was characterized by fragmentation in that the various directives focused on different (mainly user-specific) water quality and water quantity objectives. Furthermore, the EU water policy did not only suffer from weak implementation records in a couple of member states but also from negative impacts stemming from the incongruence of water management institutions and the water bodies. Also, other EU sector politics had negative effects (e.g., agriculture and transport), which were partly in contradiction with the water protection objectives. Consequently, the complex interactions between aquatic and terrestrial ecosystems, surface waters and groundwaters, as well as freshwater bodies and the marine environment were not sufficiently taken into account. In particular, regulations relating to fluvial dynamics and the interaction between water and land were, if they existed at all, selective and insufficient in respect to the achievement of environmental objectives. Compared to this traditional approach, the WFD clearly represents a substantial progress towards integration and comprehensive management of water and water-dependent natural resources and ecosystems. On the level of the general objectives, the WFD aims to build up a framework for the protection of not only inland surface waters, but also transitional waters, coastal waters, and ground-water (Art 1). Furthermore, Art 1 calls for the protection and enhancement of the status of aquatic ecosystems and, with regard to their water needs, terrestrial ecosystems, and wetlands directly depend-31 ent on the aquatic ecosystems. Equally, the WFD explicitly aims to contribute to the general protection of the marine environment and to achieve the concrete objectives of international agreements aiming to prevent pollution of the marine environment. Furthermore, the definition of 'good status,' which is the overall objective of the directive for the surface waters, can only be reached if member states take on an integrated perspective that respects interdependencies in water use and protection of the ecosystems.

Against this background, the WFD illustrates the application of Principle 3 of the EsA in several ways. First, the directive focuses on the whole river basin as its functional unit. Accordingly the environmental objectives, management plans, and programmes of measures should be defined or integrated on the level of the river basins. Therefore, the directive will give water managers (institutional and instrumental) means to protect aquatic ecosystems in a way that takes into account effects of their activities on other aquatic ecosystems within the same catchment. In particular, the WFD offers room to deal with upstream-downstream externalities and the key relationships among the elements of the hydrological networks not only within the borders of the member states but also between the riparian states of the respective water bodies. The WFD, however, offers only few provisions and guidance on how expectable upstream-downstream conflicts between or within member states can be solved, e.g. in the context of the programmes of measures and the distribution of costs.

Second, the directive explicitly acknowledges the interaction and interdependencies between the different 'water types' (surface waters, transitional waters, coastal waters, and groundwater) in terms of quality and quantity. For instance, river basin management plans should not only reflect surface water conditions, rather quality and quantity objectives relating to groundwater should also play an important role. In addition, the directive shows strong sensitivity to the protection of the marine environment, particularly via the control of the use (respectively the phasing-out) of hazardous substances which are to be determined according to the provisions of Art 16 of the WFD. The required reduction of diffuse sources explicitly aims at a protection of the marine environment from eutrophication. Also, the directive encompasses the full implementation of a couple of previous water directives inter alia the Nitrates Directive.

Third, the WFD deals not only with water quality but also with the quality of freshwater ecosystems in terms of aquatic habitats (morphology of rivers, meandering, etc.). This aspect is closely related to the protection or restoration of wetlands, which are explicitly mentioned in Art 1 of the WFD. However, the directive neither includes a definition of wetlands nor does it imply specific objectives or recommendations for their protection. However, the WFD provisions in relation to wetlands (and floodplains) offer plenty of room for a strengthening of wetland protection in the EU member states.Examples are the obligations : to protect surface waters according to Art 4 (1a) which applies to 'open water' wetlands as far as they are identified as water bodies, to reach good hydro-morphological conditions of surface water bodies because the quality elements of a surface water body include the structure and conditions of riparian, lakeshore or inter-tidal zones, and hence the conditions of any wetlands encompassed by these zones. Equally, member states are required under Art (11 (3)) to establish measures to control and mitigate modifications to the structure and the condition of the riparian zones, to protect or restore wetlands because of their contribution to the achievement of the environmental objectives relating to surface waters (good ecological status, good ecological potential, etc.) or groundwater, and to protect wetlands according to the EU Habitats Directive (92/43/EEC) and the Birds Directive (79/409/EEC). Against this background, the directive offers plenty of opportunities to better protect and respect the connectivity between the river channels and the floodplains in dynamic fluvial systems and complex terrestrial-hydrological boundary interactions.

Despite the various requirements relating to wetlands and the interaction of land-use and water management, the text of the directive appears somewhat unspecified and vague. The reason is most likely that some member states feared the additional cost implications of including wetlands in the water management objectives in a more stringent and detailed manner. However, the guidance document on wetland management which was published only recently, offers some definitions and clarifications which show the potential to raise the profile of wetlands in the implementation of the WFD, even if some NGOs still judge the state of discussion as 'minimalist'.

Summing-up: the WFD broadly considers Principle 3 of the WFD even if some requirements remain unclear concerning their implications on the implementation level, e.g. the conciliation of diverging interests in the context of upstream-downstream inter-linkages. In any case, the WFD represents an important progress compared with the rather fragmented approaches of the previous regulations.

Principle 4

Recognizing potential gains from management, there is usually a need to understand and manage the ecosystem in an economic context. Any such ecosystem-management programme should:

(a) Reduce those market distortions that adversely affect biological diversity;

(b) Align incentives to promote biodiversity conservation and sustainable use;

(c) Internalise costs and benefits in the given ecosystem to the extent feasible.

Rationale: The greatest threat to biological diversity lies in its replacement by alternative systems of land use. This often arises through market distortions, which undervalue natural systems and populations and provide perverse incentives and subsidies to favour the conversion of land to less diverse systems. Often those who benefit from conservation do not pay the costs associated with conservation and, similarly, those who generate environmental costs (e.g. pollution) escape responsibility. Alignment of incentives allows those who control the resource to benefit and ensures that those who generate environmental costs will pay.

It is not an overstatement to say that the WFD is not only ambitious on the level of environmental objectives but that the directive represents an important "economic" approach to water management in parallel. The directive aims at the integration of a wider range of measures, including pricing and economic and financial instruments, to achieve the environmental objectives with a common management approach.

Second, the polluter pays–principle (or user pays-principle) shall be regarded and Art 9 of the WFD requires to follow account the principle of the recovery of costs of water services, including environmental and resource costs. More precisely, member states shall ensure by 2010 that water pricing policies provide adequate incentives for users to use water resources efficiently, and thereby contribute to the environmental objectives of the directive. Furthermore, an adequate contribution by the different water users, at least disaggregated into industry, household, and agricultural users, to the recovery of costs is expected. In applying cost-recovery principles, member states may have to regard to, inter alia, the social, environmental, and economic effects involved.

Third, economic appraisal methods must be used to guide water resource management decisions according to the requirements of Art 11, which describes the necessary steps to develop the programmes of measures. Inter alia, member states are required to establish the programmes with references to the economic analyses (Art 5) – note that the economic analyses include judgments about the most cost effective combination of measures – and to develop measures deemed appropriate for the purpose of achieving cost recovery.

Fifth, the economic analyses play a role in the designation of surface waters as heavily modified water bodies – note that this designation involves less stringent environmental objectives – because, according to Art 4 (3), disproportionate costs may justify this decision. Equally, disproportionate costs may be accepted as a reason for an extension of the deadline for meeting the environmental objectives (Art 4 (4)) and may justify the establishment of less stringent objectives according to procedures and prerequisites prescribed in Art 4 (5).

Against the background of these broad requirements, the WFD includes a highly challenging economic approach. From the viewpoint of the member states, several requirements are not easy to comply with or are in contrast to the hitherto existing practice. Here we can distinguish those requirements that demand the application (or even development) of new scientific methods (e.g., for the forecast of water availability and use, the economic valuation of water uses, and the determination of environmental and resource costs) from the implementation of new water pricing strategies or new incentive measures. Whereas the former is an important challenge for the scientific economic community and some methodological problems appear far from being fully solved (valuation of environmental costs via the application of non-market based methods, handling of uncertainties, etc.), the latter relates to the effective implementation of new water pricing strategies and incentives schemes.

In this context, it is important to note that water prices in most EU member states do not fully cover the costs and this fact probably explains why the regulations of the WFD relating to water pricing were heavily disputed during the decision-making process. As a general rule, pricing structures for municipal and industrial water services in Europe increasingly reflect the full economic costs (infrastructure, operation, and maintenance) of providing the services, but still with an important variability between the

EU member states. For instance, some member states still do not (or only partly) integrate capital costs in the calculation of the water prices for industrial and domestic water users. In addition agricultural water use – primarily for irrigation – remains heavily subsidized, which surely encourages inefficient water use. Furthermore, environmental and resource costs are better reflected in water prices where water abstraction and/or pollution charges are levied. Yet, these instruments are not common in all member states and even in those states where they exist, current rates are generally too low to cause measurable incentive effects. In addition, there is a broad lack of incentives schemes for the agricultural sector and all other land-use related aspects of water management. There is no debate yet on the economic analysis of inland fisheries (commercial and recreational) beyond an assessment of the regional importance as an economic sector. An economic analysis of this "water" use similar to analyses of abstraction and discharges is still wanted. On the contrary, investments in aquaculture and hatcheries are subsidised nationally and by the EU.

Concerning the implementation timetable, the major issues to be resolved before 2010 will be whether quantitative and pollution-related water prices will be required and what degree of subsidies (probably only for a share of capital costs) will be allowed. This will constitute a major political problem for agricultural uses in the southern member countries. Because the Programme of Measures (PoM) will probably cause additional mitigation costs for a number of users, it will be unlikely that there will be a recovery of the remaining environmental costs beyond the current level. There is a debate among economists (e.g., in Germany and the UK) on the recovery of user costs by establishing tradable water rights but with little positive response from policy actors. However, abstraction charges for bulk raw water to recover administrative costs are in place.

Against the background of the current water pricing practices in most EU member states, it is not surprising that there are several wordings in the directive – in particular the introduction of the possibility not to comply with the cost recovery principle for social reasons – that will considerably soften the effect of the requirement on the national implementation level. Germany is one of the member countries where the level of cost recovery is close to full cost, but it is understood that the existing application of the polluter-pays principle and the existing (relatively low) wastewater effluent fee already are an adequate implementation of the requirements of Art 9 (paragraphs 1, 2 and 3). Furthermore, there still are different perceptions regarding the application range of several requirements. One of the most debated issues was the definition of water services, which fall under the cost recovery requirement. In this context, it is particularly controversial whether – and if so, how – infrastructure measures would fall under this requirement. Albeit the guidance document of the WATECO group could somewhat clarify these issues, the WFD leaves much space for national interpretations.

Summing-up, because the WFD demands an economic analysis of river basins and, in principle, cost-recovery pricing, as well as the consideration of further economic incentive schemes, the consideration of Principle 4 is largely assured. However, it remains to be seen whether, and if so to what extent, EU member states will change the parts of their current water pricing approaches that are clearly at odds with the requirements of the WFD.

Principle 5

Conservation of ecosystem structure and functioning, in order to maintain ecosystem services, should be a priority target of the ecosystem approach.

Rationale: Ecosystem functioning and resilience depends on a dynamic relationship within species, among species and between species and their abiotic environment, as well as the physical and chemical interactions within the environment. The conservation and, where appropriate, restoration of these interactions and processes is of greater significant for the long-term maintenance of biological diversity than simply protection of species.

The water policy of the European Community has developed into a patchwork of directives with different instruments applied, pollutants covered, and environmental objectives pursued. The WFD constitutes a major reversal of the past piecemeal approach as it makes the functioning of aquatic ecosystems a priority of its approach. The traditional concern with chemical water quality for particular water uses has been replaced by a complex process of defining objectives in reference to the natural condi-36
tions of water bodies and to use chemical, morphological, and biological criteria to specify the objective. This process fits into a change of scientific and public perceptions of water issues towards viewing aquatic ecosystems as a part of a cycle that includes variability and interaction with terrestrial ecosystems. National water policies were broadened to encompass the restoration of river systems, including efforts to restore single species habitats, but remained limited due to financial restrictions. The high

frequency of flooding in the 1990s contributed to a better understanding of the role of wetlands and floodplains.

Compared to this background, the overall objectives of the WFD, to protect and restore ecological quality, which are defined in the Preamble para. 11 correspond to this principle without using its wording. Particularly relevant are Art 1, Art 4, Art 5, and Art 16. Further, different guidance documents, such as the guidance documents addressing wetlands and heavily modified water bodies, also contribute to guiding the application of this principle.

The Art 1 'Purpose' describes why it is necessary to establish the Water Framework Directive. The aim is to protect and improve the aquatic environment for a sustainable water use. An assumption required for protection is a certain level of ecological quality. A claim of Principle 5 (implementation guidelines) is to take account of or to minimize risks to ecosystem functions and structures. Art 5 calls for a general review of environmental impacts resulting from human activities at the level of the river basin district. In this context Art 16 calls for the European Community to adopt specific measures against pollution of water by pollutants presenting a significant risk to the aquatic environment and to progressively phase out the discharges of priority substances, based on an established risk assessment procedure. As the European community has agreed on the procedure for a few pollutants so far, the WFD requires the Commission to submit a list of priority substances. A consensus on these pollutants could not be reached during the formation of the WFD. Thus, the potential risk which these hazardous pollutants pose has not been addressed at the European level.

Further, member states must protect and enhance all artificial and heavily modified water bodies with the target of good ecological potential and good surface water chemical status. Equally, member states must prevent or limit the input of pollutants into groundwater and prevent the deterioration of the status of all groundwater bodies. In addition, member states shall protect, enhance and restore all groundwater bodies and ensure a balance between abstraction and charge of groundwater, with the aim of achieving good groundwater status.

The ecological status of the surface water bodies will be classified with biological quality, hydro-morphological quality, and physico-chemical quality elements. This shows that not only one aspect determines ecological status, but rather the interactions of the ecosystem. The sum of all quality elements indicates the ecological quality of surface water bodies. The assessment of the ecological status classification includes the aim of Principle 5, because this principle says ecosystem functioning is an inter-relationship within and among species, between species and their abiotic environment, and includes physical and chemical interactions. In this context, it is also important to emphasise the link between the WFD and the EU-legislation for protected areas. For instance, Art 4 (1c) requires the member states to comply with any standards and objectives for protected areas under community legislation within 15 years while Art 6 requires the establishment of a register of protected areas in all river basins. The most relevant effect of these provisions is the integration of the protected areas on the level of water management objectives and of the level of RBM plans. This is clearly a step forward compared to the previous provisions where EU water directives and nature protection directives were barely integrated.

To achieve good ecological quality in river basins is it particularly necessary to integrate wetlands, because they are an essential link between surface and ground water bodies and between water and land. They are also a part of the water environment. The WFD does not use the words "to protect wetlands" as a specific objective, but it formulates the equivalent in protecting aquatic ecosystems and, with regard to their water needs, terrestrial ecosystems and wetlands directly depending on the aquatic ecosystems (Art 1 (a)). During the WFD implementation, the guidance document on 'Wetlands' has the function of ensuring that member states take into account the links between the ecological quality objectives and the functions and values of wetlands.

Furthermore, the guidance document 'Identifying and Designating Heavily Modified and Artificial Water Bodies' (HMWB) is important. If a water body is designated as heavily modified, the ecological status must only have good potential. This is a result of the hydro-morphological changes which are reason for HMWB designation. A river with good ecological potential implies lower biological status, yet still a good chemical status.

The specific strategy and policy of the WFD is the holistic mindset. Issues are not only about reduction of water pollution but also about integrating surface and ground water bodies with their aquatic and terrestrial ecosystems, as well as their transitional and coastal waters. However, wetlands as an important component of aquatic ecosystems are not explicitly mentioned in substantive articles such as Art 4. However, the planning procedure according to WFD follows Principle 5 quite well: The assessment of the current status covers the situation of water bodies. The interaction of species and their abiotic environment is covered in the definition of the reference conditions, in the classification of the current status, and in the analysis of the impact of human pressures. The identification of water bodies at risk of not complying with good status will initiate the search for restoration measures of components currently not reaching good ecological status.

It remains an open question whether all member states will fully accept the priority of the objective to conserve ecosystem structure and functioning. As mentioned before, there are national differences regarding the priority of environmental objectives in general and the ecosystem goals for rivers and lakes in particular. There is a North-South difference on the relative importance of agricultural water use and protecting freshwater ecosystems for ecological purposes. This difference became obvious during the negotiations of the WFD. The options that the WFD provides to use the exemption of HMWB or to delay the implementation will be used quite differently, as the implementation of the other water-related directives shows.

Summing-up, the WFD corresponds with Principle 5 quite well. The emphasis is clearly on structures, processes, and the dynamics of freshwater ecosystems but not on individual species. The focus on the conservation of species (and equivalently on habitats) is the core of two earlier directives, the Habitat Directive and the Birds Directive. The link to the dynamic interaction with wetlands and flood-plains is rather indirect, and the appropriate consequences for land use are not integrated into the WFD. Consequently, recognition of all functions and roles played by floodplains and wetlands (e.g. groundwater recharge, protection of water quality) is not fully assured in the implementation process of the WFD.

Principle 6

Ecosystems must be managed within the limits of their functioning.

Rationale: In considering the likelihood or ease of attaining the management objectives, attention should be given to the environmental conditions that limit natural productivity, ecosystem structure, functioning and diversity. The limits to ecosystem functioning may be affected to different degrees by temporary, unpredictable or artificially maintained conditions and, accordingly, management should be appropriately cautious.

The aim of this principle is not explicitly mentioned in the Water Framework Directive, but the approach is based on an understanding of the limits of aquatic ecosystems as the requirements of aquatic species and the functioning of their habitats are explicitly set. The objectives for inland and coastal surface waters, as well as for protected areas should be used to limit human use of water bodies. The common understanding within the European Community is that some of its waters have been used beyond their limits, particularly with respect to groundwater pollution and physical modification of surface waters. Water abstractions have been reduced in Western Europe, but the Mediterranean area shows signs of groundwater overexploitation, with a high degree of agricultural water use and plans for further expansion with financial support from EU funds. Nitrates and pesticides are often a problem in groundwater used for drinking water. The latter are still a problem in surface waters whereas there are successes in reducing eutrophication (EEA 2003c). The surface waters have been modified to protect against floods and to be used for shipping. River beds have been used for mining (sand & gravel). Commercially used fish stocks of inland waters that are dependent on natural habitats have become depleted or overfished in wide stretches of water. Viewing the rivers together with their floodplains has led to a number of cases of restoration efforts. In this sense, the scientific community and environmental groups share a perspective that we are already beyond the limites of functioning for many aquatic ecosystems, at least for the aquatic ecosystems in Western and Southern Europe.

With respect to the effects of pollution, the precautionary principle has been given high priority in Art 10, Art 13 and Art 16. In accordance with Art 16, a new directive has been proposed by the Commission in 2003.

Central to this principle is the understanding of the limits of ecosystem functioning and the effects of various human uses on the ability of ecosystems to deliver goods and services. Although the water ecosystems in Europe are well researched, there are still problems in understanding and knowing their ecosystem limits. However, the WFD reflects this consideration as it establishes a multiphase process for the implementation of an River Basin Management plan (RBM plan), which requires a separate phase to assess the status of the water ecosystem concerned, a review of the impact of human activities in Art 5, and a monitoring system staged according to the degree of uncertainty in Art 8. In addition, the WFD includes obligations for the European Parliament and the Council to adopt measures aiming for the progressive reduction and, for the cessation or phasing-out of discharges or emissions of priority hazardous substances (Art 16 (1)). Here, a conflict among member states about the content of the list of these substances has prevented a regulation within the WFD.

The regulation contained in Art 4 (1a iii) for heavily modified surface water bodies is a mechanism to identify potentially unsustainable water uses with respect to ecological status. Art 4 (3 a and b) provides a mechanism to test whether and to what extent the HMWB can be restored or modified via changes to the hydro-morphological characteristics of the water body. It is a relatively explicit procedure to evaluate alternatives with an economic reasoning. If properly followed, it amounts to a reversal of the burden of proof on what has to change: Now the degraded status quo has to be justified, not the environmental objective any more. At the same time, there is suspicion that the regulation might be used by member states as a backdoor to avoid the stronger environmental objective of good ecological status. It will be difficult to assess the actual outcome because the conditions will vary among the member countries and the methods to designate the water bodies will be different according to the particular circumstances as can be seen in the pilot river basins (Kampa & Hansen 2004).

The need for adaptive management strategies is a central consequence of Principle 6. Besides understanding ecosystem limits and avoiding adverse impacts, adaptive management implies a focus on active learning through environmental assessment and monitoring. In the implementation process of the WFDthis will be ecosystem–based as the assessments of the current status, the analysis of the pressures and impacts, and the monitoring will be performed on the level of river basins and will be included in the required river basin management plans.

The ecological guidance documents, the monitoring guidance document, and the guidance document 'Analysis of pressures and impacts' (No 3) are all relevant to this principle because they help to identify and to understand the limits of water ecosystems functioning and to develop an adaptive management approach. The WFD includes a much wider range of pressures on aquatic ecosystems in contrast to previous EU water legislations which focused mostly on point discharges. The WFD and guidance document No 3 on pressures and impacts emphasise to identify all factors that affect the ecological quality, including land-use patterns, morphological change to water bodies and diffuse pollution.

Summing-up: This principle is reflected to a considerable extent in the WFD, particularly with respect to the need to understand ecosystem functioning and to learn about assessing ecological and chemical status of the water bodies. As this is the first step in the planning process for river basins in the EU, it currently implies considerable administrative efforts in the member states. Despite the expectable shortcomings on the implementation level and some little ambitious specifications in the hitherto prepared guidance documents, the WFD can be described as an adaptive management strategy. However, the WFD does not embody the principle to its fullest extent possible as it allows several exceptions in the formulation of its environmental objectives, the most important being the designation of water body as heavily modified. However, in such cases it foresees restoration measures to be included in the RBM plan. This could be seen as an important step in the direction of managing the aquatic ecosystems of Europe within the limits of their functioning.

Principle 7

The ecosystem approach should be undertaken at the appropriate spatial and temporal scales.

Rationale: The approach should be bounded by spatial and temporal scales that are appropriate to the objectives. Boundaries for management will be defined operationally by users, managers, scientists and indigenous and local peoples. Connectivity between areas should be promoted where necessary. The ecosystem approach is based upon the hierarchical nature of biological diversity characterized by the interaction and integration of genes, species and ecosystems.

Temporal Scales

In general, the factor time is an important parameter in water management. For instance, seasonal patterns of rainfall and water availability shape the regional approaches to water use (e.g., navigation, and agriculture) and water management. The same is true for periodic events such as flooding and the respective protection efforts. From an ecological point of view, river systems are highly dynamic systems because the water flow regime (in connection with sediment flow and the general flooding regime) continually changes the natural conditions and the ecosystem's development and qualities. Not surprisingly, the ecological objectives of many river restoration projects emphasize the long-term effects. For instance, the development of an alluvial forest is a natural, long-term process. In river systems, very different time scales are important ranging from a couple of days to those processes that occur in intervals of up to several decades. As a background variable, expected long-term climate changes increasingly attract attention and will influence the management decisions in the years to come because of the potential effects on the general water availability and the expected accentuation of flood and drought events.

Another dimension of time involved in water management is the path dependency of water infrastructures and the institutional dimension of water management. Technical infrastructure systems that provide basic services such as drinking water supply and/or waste water collection and treatment are very capital intensive and therefore can not easily be changed in the short-term for economic reasons. Rather, the investment cycles are characterised by time intervals amounting to several decades with the consequence that some water management options, although potentially desirable in the short-term, are economically realistic only in the very long-term. The same is true for the path dependency and persistence stemming from the interaction of settlement structures and water management. Because of the past settlement activities in floodplains, restoration of their ecological functions in urban areas is regularly hampered by current forms of land-use that can not be changed in the short-term unless enormous economic obstacles and political resistance can be overcome.

Despite these obvious time dimensions of water management, the WFD does only scarcely recognise time as an explicit parameter. The explicit time dimension of the WFD is the schedule for the implementation of the directive and the deadline for the achievement of the ecological objectives (good status, etc.), which is 15 years after the date of entry into force. In addition, the WFD explicitly demands a process of review and updating the PoM every six years thereafter. However, there is a rather implicit acknowledgement of time in the definition of the environmental objectives and the PoM. While some objectives and measures show a rather short-term character, others can only be realised in the long-term. Clearly, measures aiming to restore fluvial dynamics in many rivers can only be realised in several decades due to the aforementioned economic reasons and long-term ecological processes.

In several cases, the pursuance of long-term objectives (e.g., the restoration of floodplain ecosystems for flood control purposes) might be at odds with potential short-terms gains and benefits that result from an intensive economic use of the areas (e.g., building of residential or industrial complexes on floodplain areas. In this context, the implementation of the WFD might change respective land-use decisions in favour of the recognition of long-term objectives. An explicit consideration of lags-effects that regularly characterise ecosystem processes, however, can not be found in the text of the directive.

Spatial Scales

As already mentioned, the directive is a deliberate effort to put the water policy in the EU on the appropriate spatial scale. Crucial requirements can be found in the Articles 2, 3, 4, 5, 8, and Art 14. The WFD develops a nested approach to spatial scales and it will demand at least from some member states, new institutional mechanisms. It requires the member states to designate river basins and coastal waters, to form river basin districts, and to assign the river basins and coastal waters to the nearest river basin district (Art 3) with the corresponding definitions in Art 2.

The WFD uses a nested approach to the spatial scale of water ecosystems as well as water management. A starting point is the river basin and its assigned groundwater, and surface, transitional, and coastal waters. A water body is defined as a distinct part of these water categories, which differ according to environmental status (Art 2 paragraphs 8, 9, 10, and 12). The WFD takes this as a starting point in Art 4 for achieving a good ecological water status. The guidance document 'Identification of Water Bodies' (No 2) describes how to define a water body (European Communities 2003b). The boundaries are defined on a scientific basis for the water bodies and the river basins. Here, there is no stakeholder involvement. Art 3 (1) provides a pragmatic approach to the combination of river basins and river basin districts as it distinguishes between small and large river basins and it allows for the combination of small river basins to be assigned to a district. The same procedure is applied to the definition of groundwater bodies and their assignment to RBD. The WFD leaves the organisation of large river basins up to the member states because the member states could not agree on a common institutional structure which would have required constitutional changes in the federal states (e.g.,

Germany). The question of the optimal level of decision-making remains unanswered (Compare Principle 2).

Summing-up: Previously, water management in a number of member states stopped on intra-national administrative borders and international borders. Now with the WFD, all the member states have to manage their water ecosystems in river basin districts, even if they extend across administrative regional or national borders. However, the WFD does not define the spatial water management structure; it leaves it up to the member states. It only requires the member states to assign transboundary river basins to international river basin districts (Art 3 (3)) and obliges the member states to produce a single international river basin management plan (Art 13 (2)) if the basin fully falls within EU territory. If the river basin extends beyond the EU territory, it obliges the member states to search for an international RBM plan (Art 13 (3)). The analysis of issues, pressures, and status will occur on the appropriate spatial scale as well as the setting of the objectives and the implementation of measures and plans. There is no explicit incorporation of the appropriate temporal scale, only a tight schedule for implementation. The provisions for repeating of the planning cycle after 2015 can be viewed as an implicit recognition of the long-term character of the restoration challenges lying ahead for European aquatic ecosystems.

Principle 8

Recognizing the varying temporal scales and lag-effects that characterize ecosystem processes, objectives for ecosystem management should be set for the long term.

Rationale: Ecosystem processes are characterized by varying temporal scales and lag-effects. This inherently conflicts with the tendency to favour short-term gains and benefits over future ones.

Because there is considerable overlap with Principle 7, see the discussion of principle 8 under Principle 7.

Principle 9

Management must recognize that change is inevitable.

Rationale: Ecosystem change, including species composition and population abundance. Hence, management should adapt to the changes. Apart from their inherent dynamics of change, ecosystems are beset by a complex of uncertainties and potential “surprises” in the human, biological and environmental realms. Traditional disturbance regimes may be important for ecosystem structure and functioning, and may need to be maintained or restored. The ecosystem approach must utilize adaptive management in order to anticipate and cater for such changes and events and should be cautious in making any decision that may foreclose options, but, at the same time, consider mitigating actions to cope with long-term changes such as climate change.

Changes in European freshwater ecosystems occurred naturally in the past quite frequently because of the dynamics of morphological parameters of rivers. These changes occurred with considerable variation on a spatial and temporal scale, and altered habitat structure and species composition. This process has been modified due to anthropogenic changes of land use and due to human-made modifications of rivers and river beds over the last 200 years. The expected changes in the global climate regime will alter precipitation and runoff regimes in Europe, modifying river morphology. The development of river engineering in Europe resulted partially from the desire to take advantage of the use potential of rivers and to control the vagaries of rivers, especially the effects of floods on settlements. The current scientific and political debate on watershed management focuses on a change of direction by creating more space for floodplains.

Because the WFD focuses on the restoration or mitigation of man-made degradation of the EU water ecosystems, it does not explicitly refer to the inevitability of change in ecosystems as a conceptual basis for its management approach. The WFD uses the natural condition of its water bodies as a reference for the quality objectives to be pursued and it implicitly recognizes the need to adapt to natural changes. It deals with changes arising from socio-economic factors as well, which more explicitly determine the ability and willingness of member countries to reach the stated objectives. The topic of flood and drought management is mentioned only in Art 1 as the fifth purpose of the WFD, but it is not specified further at another place in the directive.

The WFD deals explicitly with changes that can be expected from human pressures. It requires the member states in Art 5 to review the impact of human activities on the status of surface waters and groundwater. The WFD also specifies the types of pressures to be identified and estimated. This can be implemented by developing scenarios about human pressures (until 2015) and then modelling their impacts. This is a plausible way of dealing with socio-economic uncertainties. The exemptions listed in Art 4 may be seen as another element of adaptive management, but they are exemptions from the obligation of achieving good water status. Two of them have socio-economic reasons: the heavy modification of the water body (Art 4 (3)) and technical and economic feasibility (Art 4 (4)). Only one exemption refers to natural changes (Art 4 (6)). A third element of adaptive management is the monitoring requirements in Art 8, which are designed to achieve the desired objective of good water status. Natural ecosystem changes will be registered by the monitoring programs as well, but they will be mostly designed to grasp the need for the Programme of Measures and then to monitor their effect as required by Art 11 (5). The flexibility of the WFD can be seen in its reiteration of obligations of member states to repeat certain steps after they have been performed for the first time. Usually, the update period is set for six years.

The guidance document on the "Planning Process" provides an overview of the planning approach of the WFD beyond the text of the directive. It reiterates the cyclic nature of the planning process as a general principle and relates it to the schedule set by the directive. It emphasizes that there are points for adaptation, particularly in the early phase of implementing the WFD. The first step, the assessment of the current status includes a preliminary analysis of the gap between status and objective and a preliminary designation as a heavily modified water body. Then, there are steps of risk assessment, which are based on the results of the monitoring activities in the middle of the implementation schedule. These assessments mainly focus on the goal of achieving good water status, and less on identifying natural perturbations. For the purposes of the WFD, it becomes important to take changes of the socio-economic drivers and their impacts into account. The identification of significant water issues and the development of a baseline scenario can illustrate the economic dynamics in the river basin.

All these points are analysed in more detail in the relevant guidance documents: the guidance documents on the reference conditions and ecological status for inland surface waters (No 10) on the typology, reference conditions, and classification systems for transitional and coastal waters (No 5), on the identification and designation of heavily modified and artificial water bodies (No 4), on economics and the environment (No 1), and monitoring (No 7). The guidance document about the analysis of pressures and impacts (No 3) provides more information on the analysis of pressures. The references for changes are pressures which influence well known sources of pollution and other alterations of water bodies. This aspect is explicit in the guidance document, developed on the "statistical aspects of the identification of groundwater pollution trends and aggregation of monitoring results," because one sub-objective for groundwater has to be operationalised "to reverse any significant and sustained upward trend in the concentration of any pollutant". In all these guidance documents, the preliminary nature of the first assessment is emphasized, but it is expected that this assessment eventually will lead to decisions.

The visible implementation efforts on a national and RBD scale show that the assessment of the current status of the water bodies is subject to a number of uncertainties, particularly with respect to the ecological status of surface waters and the development of pollutant concentrations in groundwater on the scale and complete coverage required by the WFD. The available documents in the German state of Hesse show a high share of uncertain assessment. This it due to the lack of available information rather than to the lack of understanding of a complex ecosystem.

Summing-up, the WFD implicitly provides for the recognition of the inherent change in freshwater ecosystems by forcing water managers to take a broader view of surface and groundwater status and assessing the status of their waters in a comprehensive manner. The term "status" is not associated with a dynamic perspective, but based on the status quo. A projection of pressure-induced changes and measures to improve the situation have to be made in order for the river basin management plan to be effective. This can only be accomplished with an understanding of the dynamic nature of the ecosystem concerned. The WFD makes the anticipation of human-induced changes a required task by demanding the analysis of pressures and impacts. The planning process, particularly the design of the Programme of Measures, itself will create additional uncertainty demanding flexibility from all parties involved.

Principle 10

The ecosystem approach should seek the appropriate balance between, and integration of, conservation and use of biological diversity.

Rationale: Biological diversity is critical both for its intrinsic value and because of the key role it plays in providing the ecosystem and other services upon which we all ultimately depend. There has been a tendency in the past to manage components of biological diversity either as protected or non-protected.[1] There is a need for a shift to more flexible situations, where conservation and use are seen in context and the full range of measures is applied in a continuum from strictly protected to human-made ecosystems.

Biological diversity in European freshwater ecosystems is low in a global comparison, but the available components (e.g. fish, water birds, reeds) are heavily used and their composition and abundance has been heavily modified. At the same time, there are considerable efforts for the conservation of wetlands biodiversity and for individual species (e.g. salmon, mussels). Pollution and the modification of rivers, lakes, and coastlines over the last two centuries changed the habitats for many species considerably, so that the spectrum of usable components was reduced and modified as well. The total number of species has been reduced on a limited scale, while the genetic diversity of the species in aqua-culture is lower due to the use of hatcheries. But for several important species in aquaculture, the genetic resources reside in natural populations. Abundance of fish stocks improved as water quality and the ecological quality of rivers improved. The use of biodiversity focused on fish and shellfish consumption on a commercial scale and to a limited extent on wild waterfowl. Aquaculture has been introduced as a substitute for wild fisheries, which has reduced the level of catch from the wild. The remaining wild stocks have been overused and require restocking based on hatcheries. Recreational fishing has gained importance, but it relies to a large extent on hatcheries as well. Waterfowl hunting is of recreational importance, but national conservation laws and an EU regulation are in place to limit the use of and to protect endangered species.

Protection of biodiversity in freshwater ecosystems is scattered over regional and national nature protection regulations and supported by international conventions such as the Ramsar Convention for wetlands of international importance. In addition, the Habitat Directive of the European Union protects habitats and species of European significance. The WFD does not regulate the use of elements of biological diversity, neither for commercial nor for recreational purposes. The EU is responsible for the common fisheries policy, which focuses on marine fisheries, but covers inland fisheries as well. It supports inland fisheries by a number of instruments: e.g., product specification and standardization, price support, investment subsidies, and marketing support, which targets the market side of the fishing industry with the aim of supporting income and employment. The resource side of the industry - water, fish, shellfish stocks, sites, and fishing rights - is regulated on a national or state level in contrast to the marine part of the EU fisheries. Parallel to isolated national river restoration efforts, the reintroduction of endangered species is becoming a topic on a national level, which tends to attract public attention particularly for salmon in European rivers where they have become extinct. Here, the International Commission for the Protection of the Rhine has been particularly active to reintroduce salmon population.

The WFD does not deal with these issues because it is conceived as a directive to design the framework to protect water bodies and to use the water in a sustainable manner. It supports the conservation of biological diversity indirectly by defining its objectives via good ecological status, which includes biological elements, hydromorphological elements, and chemical and physico-chemical elements supporting the biological elements. Aquatic flora, phytoplankton, benthic invertebrate fauna, and fish fauna are to be assessed in relation to their species composition, abundance, biomass, and age structure for the four types of surface water bodies: rivers, lakes, and transitional and coastal waters. For groundwater, quantitative status and chemical status are to be assessed by determining whether there is "any significant damage to terrestrial ecosystems which depend directly on the groundwater body".

The guidance documents on reference conditions for inland waters and on coastal waters specify the criteria for the assessment further by including pressure screening criteria and by providing interpretation for the normative criteria for reference conditions and the related "good ecological status." To a limited extent, this can be seen as an attempt to define a set of biodiversity objectives in terms of taxonomic composition and abundance. This link is emphasized by an additional horizontal guidance document on wetlands, which was endorsed by the water directors in November 2003

It clarifies the relationships among the water bodies for which objectives are defined. The guidance document on wetlands also clarifies the functional role of wetlands within the hydrological cycle and the river basin ecosystem. It lists and defines the main wetland elements covered by the WFD: wetland ecosystems identified as water bodies, riparian, shore and intertidal zone quality elements of surface water bodies, the terrestrial ecosystems directly depending on groundwater bodies, Small elements of surface water connected to water bodies but not identified as water bodies Ecosystems significantly influencing the quality and quantity of water reaching surface water bodies, or surface waters connected to surface water bodies

By reiterating the environmental objectives of the WFD, by specifying the biological quality elements for each wetland/ water body, and by providing best practice examples, this guidance document provides an indirect link to biodiversity planning without ever giving it such a name. In the same style, it analyses the implications of the designation of HMWB for the role of wetlands, their role in the analysis of pressures and impacts, as well as in the programme of measures.

By explicitly including a list of protected areas, and by broadening the objectives of water management by including ecological quality (biological elements and their hydromorphological and physico-chemical prerequisites), the WFD supports the conservation side of principle 10, particularly on the level of ecosystems and species. However, it does not cover the use side of biodiversity and it does not address the balance or integration of conservation and use. The guidance document on the Analysis of Pressures and Impacts includes aquaculture as a point source and fishing and fish stocking as pressures with an impact on living resources, but does not help with tools to perform the analysis. There is also a recommendation to include an analysis of the pressures from the biological quality elements on the potential reference conditions or sites. But performing these analyses will not amount to the balancing or integration of conservation and use of biodiversity. The use side (i.e. inland fisheries) has been regulated by the EU, but mostly by the member states according to their national priorities, but this is mostly done according to sectoral objectives (competitiveness, marketing considerations, etc.). There are signs for a change, particularly on the EU-level, when the Commission decided to integrate environmental concerns into its Common fisheries policy which concerns mostly its marine fisheries, but in 2002 it developed a strategy for sustainable development of European aquaculture.

Aquaculture and inland fisheries are highly diverse and consist of a broad spectrum of species, habitats and harvesting practices in Europe, which is accompanied by a broad range of fishing rights and regulation by its member states, often on a regional level. The regulations tended to be of a sectoral focus, but in a number of cases, integration is progressing: Aquaculture, river and lake fisheries and recreational fisheries are seen together, their stocking activities are coordinated and biodiversity considerations play a role in the stocking decision as well. But a consistent integration on the basis of river basins is still lacking. There are signs that the fishing stakeholders are increasing their attention to the implementation of the WFD on a regional level, but their main concern is still sectoral. The only exception here is on a species level:

The development of an action plan for the management of the European eel by the European Commission can be seen as a species-specific approach to the integration of conservation and sustainable use. It remains an open question whether the path towards integration will be limited to a species-specific approach or whether the political demands for more conservation and the potential opposition by the fisheries stakeholders will be balanced by the upcoming planning process of the WFD.

Summing-up: While the WFD mentions sustainable quantitative water management as a key objective of the directive, the balance between use and conservation of biodiversity elements and services is not explicitly treated. However, there is a certain indirect consideration because of the interaction of the WFD with EU nature protection directives, and because of the provisions for the determination of environmental objectives.

Principle 11

The ecosystem approach should consider all forms of relevant information, including scientific and indigenous and local knowledge, innovations and practices.

Rationale: Information from all sources is critical to arriving at effective ecosystem management strategies. A much better knowledge of ecosystem functioning and the impact of human use are desirable. All relevant information from any concerned area should be shared with all stakeholders and actors, taking into account, inter alia, any decision to be taken under Article 8(j) of the Convention on Biological Diversity. Assumptions behind proposed management decisions should be made explicit and checked against available knowledge and views of stakeholders.

Depending on the national starting point, quantitative knowledge about the status and use of surface and groundwater including ecological status is available within governmental agencies, but not always well shared among the agencies. Quite often information about water availability, quality, and use is kept in agencies separate from other agencies that have information about biological elements, biodi-49 versity use, and water-relevant land use. Considerable knowledge is available at governmental and university research institutes. In some member states, there are segments of water use and use of biodiversity elements which are self-governed, for example in recreation and inland fishing, in which cases information is not available in a comprehensive manner to central governmental agencies.

The WFD is a consistent effort to bring all relevant information from all sources to the appropriate level of decision-making, in this case the River Basin District. Yet, as a European instrument, it does not relate to indigenous sources of knowledge that are not part of the established network of know-how. The WFD asks the member states to encourage the active involvement of all interested parties in the implementation of the directive (Art 14 (1)). More specific requirements relate to the duty to publish the timetable and work programme for the production of the plan, an interim overview of the significant water management issues, and a draft of the river basin management plan, as well as to make them available for comments to the public and the users (Art 14 (1)).

The guidance documents that are relevant to the first step of the assessment of the current status, pressures, impacts, and the economic analysis of current water use are all based in the relevant sciences. The definition of ecological status and the calibration of the testing at the various sites is a pan-European effort of the relevant scientific organizations involved. This process is basically open, but it is not of interest to the wider public or the media. It is basically a governmental inter-organisational effort to search for the necessary data, provide them in an appropriate spatial format, and make them manageable on the basis of electronic data management. In Germany, the challenge of coordination is higher then in centralised states, because data and information systems have to be made consistent between the different Laender administrations.

The national implementation of public participation has led, so far, to a broad presentation of the regulations, and the publication of the planning process, and schedule on the websites of the authorities in the member countries. In Germany, one NGO with federal funding has tried to provide a critical Internet-based forum to make the implementation a topic among the local organizations of environmentalists.

A second element of public information strategies has been the establishment of advisory councils, where they did not exist before as in Germany. The purpose of these advisory councils is to represent stakeholders, i.e. user organizations and environmental protection organizations in the decision making process. Under different circumstances in which participatory decision-making procedures had already been established, advisory councils have been in place since several decades, for example in France and in the UK. It depends on the tradition of cooperation in each member state between governmental agencies on one side and nature protection organisations, fishery organisations, and recreational associations on the other side to what extent the latter's knowledge of water bodies and biological elements is integrated beyond the two mechanisms mentioned above. The knowledge of the non-governmental side tends to be considerable as these organisations include more than a hundred thousand individuals who have close and frequent contact with the water bodies in question. Still, there is criticism in Germany that their potential has not yet been fully utilized.

Summing-up, the WFD initiates a major effort to bring all forms of relevant information together on the level of the river basin district. As a first step, this means the collection and integration of governmental and scientific information about ecosystem functioning, status, and human pressures and their impacts. With established communication channels to non-governmental organisations, their knowledge is actually used. Although there are still gaps in information and information exchange, the WFD has led to a significant improvement and a remarkable increase in the amount of available data. With the provisions of the WFD for public participation, chances are good that the stakeholders will be more integrated into the planning process.

Principle 12

The ecosystem approach should involve all relevant sectors of society and scientific disciplines.

Rationale: Most problems of biological-diversity management are complex, with interactions, sides-effects and implications, and therefore should involve the necessary expertise and stakeholders at the local, national, regional and international level, as appropriate.

The consideration of Principle 12 is closely related to Principle 1 (participation) and Principle 11. The WFD does not explicitly have requirements for the involvement of other relevant societal sectors and scientific disciplines. However, we can express the opinion that a full implementation of all requirements of the WFD will clearly demand the active involvement of different sector policies and broad scientific support and involvement regardless of the WFD's lack of an explicit norm.

Two examples might serve as an illustration: Because diffuse pollution is mostly stemming from agricultural sources there is a clear need for an integrated perspective on both the water management instruments on the one hand and the agro-environmental regulations and financial instruments such as voluntary contracts with individual farmers on the other hand. The same is true for the envisaged renaturation and restoration of rivers. In these cases, significant progress cannot be reached without an active coordination with the sector policies such as energy, navigation, and agricultural policy. In addition, inter-sectoral planning instruments and land-use provisions in the form of regional and local land-use schemes will play an important role.

However, it is largely an open question as to how the involvement of the various sectors might be achieved. A large variety of models in the EU is expectable because, as of today, member states have very different administrative and political conditions. In this context, the requirement to encourage the active involvement of all relevant stakeholders as discussed in Principle 1 might play an important role in the years to come. Yet, many obstacles can be expected because coordination of water management with other sectors is largely a new challenge compared to the hitherto existing practices in many member states. This assessment is also relevant for the demanded active integration of water administrations and nature protection agencies. Since the directive cannot be implemented unless respective mechanisms are in place, many member states still are lacking adequate communication and coordination tools.

The involvement of the different scientific disciplines in the development of water management concepts in the EU already appears highly advanced. Clearly, the directive demands a far-reaching involvement of natural science and science-based modelling of river basins for the demanded river analysis and the determination of environmental objectives. In addition, the involvement of social sciences and economics is, at least implicitly, stipulated. Economic analysis of river basins and the determination of environmental and resource costs clearly can not be realized without the involvement of the relevant disciplines. Also, economic analysis should reflect the respective methodological state of the art. The same is true for the implementation of the public participation requirements of the WFD, which needs input from social sciences for the development of participation designs and adequate methods.

In this context, it is worth mentioning that the European Commission has financed a couple of interdisciplinary water-related research projects over the past years, some explicitly with the objective to support the implementation of the WFD. Equally, on the national level important interdisciplinary projects were carried out. Their results deliver valuable information for the implementation of the WFD. However, with respect to the communication of scientific results to management and the coordination and cooperation of various disciplines in large research projects, there still is room for improvement in light of the requirements of both the WFD and the EsA. Furthermore, and in spite of the integrated perspective of the WFD, some important aspects of the management of aquatic ecosystems (e.g. flood risk management, fisheries) are still dealt within other fora.

Summing-up, the WFD clearly pushes 'state of the art' water research including both natural sciences and social sciences disciplines. However, integration of scientific results in water management is still a matter of concern.

Implications for the Development of the EsA

The WFD is a special case of (implicitly) applying the ecosystem approach as it covers "only" one type of ecosystem, but on both national and international scales. Complete coverage of all inland waters is a legal obligation with a compulsory implementation schedule. The explicit objective is to improve the ecosystem quality within 15 years, i.e. to change the existing balance of protection and use considerably. With these characteristics, it is unique internationally and among industrialised countries. The one only similar approach is in the United States where a number of large scale river basins are being restored based on an ecosystem approach as well. But here the approach is selective and open-ended as it covers seven larger river basins with an existing regional history to think and act based on the ecosystem approach. It is open in terms of schedule, but some of them were explicit in calculating costs and in obtaining funding from the federal level to implement the necessary measures.

The WFD could serve as a case for implementing the EsA on a national or multinational level. It shows all the legal changes necessary to make it work on this scale. It illustrates the need to change the administrative and legal structures before the planning process on the level of the river basin can take place. The WFD shows the minimum of regulation to change towards the institutional fit in ecosystem management. The European Union has to deal with a variety of institutional approaches to water management in its member countries and, thus, the WFD is unspecific about the necessary changes, but the member states and their adaptation to the WFD can serve as a laboratory of approaches to any other country thinking about introducing a full-scale adoption of the EsA to inland water management. The EU member countries vary according to the basic institutional criteria for interested governments: existence and varying competences of river basin institutions, different degrees of integration of sectoral ministries, and variation in the vertical distribution of governmental authority.

A further note about the transferability of the approach of the WFD concerns the fact that there was a considerable degree of experience which a number of member states had with some of the innovative elements of the WFD: few countries have had river basin based authorities; some have a history of stakeholder involvement; some have considerable experience with the use of economic instruments and cost recovery in water service pricing; and some have experienced wetlands/ floodplain restoration efforts.

As a general rule, the WFD as an area-wide legal approach to water management will not easily be transferable to other regions in the world due to the particularities of the EU environmental policy, the specific multi-level regulation system in the EU, the particular administrative traditions, and the advanced economic development stage of most of the regions in the EU member countries. Apart from this, the WFD, as a piece of environmental legislation, emphasizes the general need to develop a coherent and far-reaching legal framework for the protection of freshwater ecosystems. Clearly, ecosystem management approaches on the project level should be accompanied and supported by a comprehensive legal framework. Against this background, the WFD might serve as an example for the development of a progressive water law whose effectiveness, strengths and weakness should be compared with the effects of recent reforms of the water law in other countries.

References

EEA (European Environment Agency) (1994): *European Rivers and Lakes, Assessment of their Environmental State*. Copenhagen.

European Communities(2001): Technical Report No 1. The EU Water Framework Directive: statistical aspects of the identification of groundwater pollution trends, and aggregation of monitoring results.Common Implementation Strategy for the Water Framework Directive (2000/60/EC).

Jones, T. (1996): The European Region. An Overview of European Wetlands. In: Hails, A.J. (ed.) *Wetlands, Biodiversity and the Ramsar Convention: the role of the Convention on Wetlands in the Conservation and Wise Use of Biodiversity*. Ramsar Convention Bureau Ministry of Environment and Forest. India.

Mostert, E. (2000): Models for River Basin Management. Experience from the Eurowater Countries. In: *Water Resources Management: Brazilian and European Trends and Approaches*. ABRH. Porto Algere. Pp. 193-207.

Smith, R.D. & Maltby, E. (2003): U*sing the Ecosystem Approach to Implement the Convention on Biological Diversity: Key Issues and Case Studie*s. IUCN. Gland, Cambridge.

WWF & EEB (2004): 'Tips and Tricks' for Water Framework Directive Implementation. *A resourcedocument for environmental NGOs on the EU guidance for the implementation of the Water Framework Directive*.

Bibliography

ADB (Asian Development Bank), (2001): *Water for All: The Water Policy of the Asian Development Bank,* Manila, Philippines.

Agnew. C. and Anderson. E. (1992) *Water resources in the arid realm* Routledge, London.

Anderson, D.M., Overpeck, J.T., and Gupta, A.K. (2002). Increase in the Asian southwest monsoon during the past four centuries. *Science 297*:596-599.

Appleberg, M. (1998). Restructuring of fish assemblages in Swedish Lakes following amelioration of acid stress through liming. *Restoration Ecology 6*:343-352.

Atkinson, D. (1995). Effects of temperature on the size of aquatic ectotherms: exceptions to the general rule. *Journal of Thermal Biology 20*(1/2):61-74.

Avila, A., Neal, C., and Terradas, J. (1996). Climate change implications for streamflow and streamwater chemistry in a Mediterranean catchment. *Journal of Hydrology 177*:99-116.

Bahls, P. (1992). The status of fish populations and management of high mountain lakes in the western United States. *Northwest Scienc*e *66*:183-193.

Brinson, M.M., and Malvarez, A.I. (2002). Temperate freshwater wetlands: types, status, and threats. *Environmental Conservation* 29(2):115-133.

Brismar, A., (1997): *Freshwater and Gender: A Policy Assessment,* Stockholm Environment Institute, Stockholm, Sweden.

Council on Environmental Quality (1995). E*nvironmental Quality*. 1994-1995 Report. Office of the White House, Washington D.C.

Covich, A. P. (1993). Water and ecosystems. In: Gleick P.H., (ed.). *Water in Crisis: A Guide to the World's Fresh Water Resources*. New York, Oxford University Press. pp. 40-55.

Daily, G.C., ed. (1997). *Nature's Services: Societal Dependence on Natural Ecosystems.* Island Press, Washington, D.C.

Dilks, D., (ed.) (1996). *Measuring Urban Sustainability: Canadian Indicators Workshop*, Toronto, 19-21 June, 1995. Ottawa, Ontario, Canadian Housing Information Centre. 73 p.

Douglas, I. (1983). *The Urban Environment*. London, Edward Arnold. 229 p.

Echeverria, J.D., P. Barrow, and R. Roos-Collins. (1989). *Rivers at Risk: The concerned citizen's guide to hydropower.* Island Press, Washington, D.C.

EEA (European Environment Agency) (1994): *European Rivers and Lakes, Assessment of their Environmental State.* Copenhagen.

Environmental Protection Agency (1998). *National Water Quality Inventory: 1996 Report to Congress.* U.S. EPA EPA841- R-97-008, Washington, D.C.

European Communities(2001): Technical Report No 1. The EU Water Framework Directive: statistical aspects of the identification of groundwater pollution trends, and aggregation of monitoring results.Common Implementation Strategy for the Water Framework Directive (2000/60/EC).

Fraser, A., M. Meybeck, and E. Ongley, 1995: *Water Quality of World Rivers,* UNEP Environment Library report no. 14, UNEP, Nairobi, Kenya.

Frederick, K. D. (1993). *Balancing Water Demands with Supplies: The Role of Management in a World of Increasing Scarcity.* World Bank Technical Paper No. 189. World Bank (Washington, DC). 72 p.

Friberg, N, et.al. (2011). Biomonitoring of Human Impacts in Freshwater Ecosystems: The Good, the Bad and the Ugly. *Advances in Ecological Research,* 1-68.

Girardet, H. (1990). The metabolism of cities. In: Cadman D., Payne G., (eds.). *The Living City: Towards a Sustainable Future.* London, Routledge. p 170-180.

Gleick, P.H. (ed.), (1993): *Water in Crisis: A Guide to the World's Freshwater Resources,* Oxford University Press, London, UK.

Goswami. A. (1998) *Troubled waters* Business India. August 10-23: 62-68.

Hahn, E. (1991). *Ecological Urban Restructuring: Theoretical Foundation and Concept for Action* No. FS II 91-402. Science Center Berlin (Berlin). 37 p.

Hamlin, C. (1990). *A Science of Impurity: Water Analysis in Nineteenth Century Britain.* Bristol, Adam Hilger. 342 p.

Jones, T. (1996): The European Region. An Overview of European Wetlands. In: Hails, A.J. (ed.) *Wetlands, Biodiversity and the Ramsar Convention: the role of the Convention on Wetlands in the Conservation and Wise Use of Biodiversity.* Ramsar Convention Bureau Ministry of Environment and Forest. India.

King, J.M, R.E. Tharme, and M.S. DeVilliers (eds.), 2000: *Environmental Flow assessments for Rivers: Manual for the Building Block Methodology,* Water Research Commission, Pretoria, South Africa.

Leathwick J.R.; Collier K.; Chadderton L. (2007) "Identifying freshwater ecosystems with nationally important natural heritage values: development of a biogeographic framework". *Science for Conservation* 274. p. 30. Department of Conservation New Zealand.

McCafferey. S.C. *Water politics and international law* in Gleick. P.H. (Ed.) Water in crisis: a guide to the World's freshwater resources. Oxford University Press. London.

McDonald, A.T. and Kay, D. (1986) *Water Resources Issues and Strategies.* Longman, London.

Mostert, E. (2000): Models for River Basin Management. Experience from the Eurowater Countries. In: *Water Resources Management: Brazilian and European Trends and Approaches.* ABRH. Porto Algere. Pp. 193-207.

Naiman, R.J., and M.G. Turner (2000). A future perspective on North America's freshwater ecosystems. *Ecological Applications* 10:958-970.

National Research Council. (1992). *Restoration of aquatic ecosystems: science, technology, and public policy.* National Academy Press, Washington, D.C.

Ricciardi, A., & Rasmussen, J. B. (1999). Extinction Rates of North American Freshwater Fauna. *Conservation Biology,* 13 (5), 1220-1222.

Smith, R.D. & Maltby, E. (2003): U*sing the Ecosystem Approach to Implement the Convention on Biological Diversity: Key Issues and Case Studie*s. IUCN. Gland, Cambridge.

Wallensteen. P. and Swain. A. (1996) *International freshwater resources: source of new conflict.* Department of Peace and Conflict Research. Uppsala University. Sweden.

Westmacott, J.R., and Burn, D.H. (1997). Climate change effects on the hydrologic regime within the Churchill-Nelson River Basin. J*ournal of Hydrology* 202:263-279.

Williamson, C.E. (1995). What role does UV-B play in freshwater ecosystems? *Limnology and Oceanograp*hy 40(2):386-392.

World Bank (1992) *World Development Report 1992*, Development and Environment, New York: Oxford University Press.

World Resources Institute (1998) *A Guide to the Global Environment* Oxford University Press, Oxford UK.

WWF & EEB (2004): 'Tips and Tricks' for Water Framework Directive Implementation. *A resourcedocument for environmental NGOs on the EU guidance for the implementation of the Water Framework Directive.*

XU, F.L., Tao, S., Dawson, R. W., Pen-gang, L., & Jun, C. (2001). Lake Ecosystem Health Assessment: Indicators and Methods. *Water Research,* 3157-3167.